CHEMICAL ENGINEERING METHODS AND TECHNOLOGY

NEW DEVELOPMENTS IN ALDEHYDES RESEARCH

CHEMICAL ENGINEERING METHODS AND TECHNOLOGY

Additional books in this series can be found on Nova's website under the Series tab.

Additional E-books in this series can be found on Nova's website under the e-book tab.

CHEMICAL ENGINEERING METHODS AND TECHNOLOGY

NEW DEVELOPMENTS IN ALDEHYDES RESEARCH

LUCA TORRIONI
AND
EMILIA PESCASSEROLI
EDITORS

nova publishers
New York

For permission to use material from this book please contact us:
Telephone 631-231-7269; Fax 631-231-8175
Web Site: http://www.novapublishers.com

Library of Congress Cataloging-in-Publication Data

ISBN: 978-1-62417-090-4

Published by Nova Science Publishers, Inc. † New York

CONTENTS

PREFACE

In this book, the authors present topical research in the study of aldehydes. Topics discussed in this compilation include the synthesis and properties of intermediates in reactions of aldehydes with P(III) chlorides; synthesis of heterocyclic compounds by interaction of aldehydes with monoterpenoids; update on aliphatic aldehydes in lipid foods; inhibition of microbial biocatalysts by biomass-derived aldehydes and methods for engineering tolerance; co-oxidation processes promoted by N-hydroxyphthalimide/ aldehyde systems; and the structure of gossypol condensation bis-product with 2-amino-4,6-dioxypyrimidine in acidic environment.

Chapter 1 – For over a century the reaction of phosphorus trichloride 1a with aldehydes 2, which was discovered by Fossek, has been known and studied. An aqueous work-up or heating of the reaction mixture provides organic compounds with a P(IV) - C bond. All attempts had failed to determine the structure of intermediate products formed at the initial stages of this intricate reaction. Page, Kabachnik and Atherton postulated phosphite structure $[(RCHClO)_nPCl_{3-n}$, n = 1, 2, 3] 3 as primary intermediates, but it was thought that no spectrometric or chemical evidence for the intermediacy of 1 could be found due to their easy addition of second aldehyde molecule and transformation into the secondary intermediate $X_2POCHROCHRCl$ 4. Phosphorus (III) chlorides 1 were found to contain HCl as a contaminant, even after multiple distillations. HCl can be removed from 1 by treating with tertiary amine 5 or alkyl vinyl ether (AVE). The synthesis of 3 and a secondary possible intermediate 4 using low temperatures and purified 1is reported. Purified 1 does not react with 2. Under the same conditions commercial 1 reacts with 2 extremely exothermically showing the HCl catalysis. In turn AVE or a small amount of a tertiary amine gives rise to a violent reaction of 2 with purified 1 showing base catalysis. It was found that the addition of one or two molecules of 2 to one P - Cl bond depends on the concentration of 5. When 1-5 % of 5 is used, the adducts3 drived from the addition of one molecule of 2 are formed. The intermediate 3 reverts to 1 and 2 on heating, thus the reactions are reversible. We suppose that 5 not only makes the carbonyl oxygen more reactive but also protects its electrophilic centre from the addition of a second aldehyde molecule. Ten primary intermediates 3 were synthesized and their structures were confirmed by ^{1}H, ^{13}C and ^{31}P NMR. It has been shown that electron acceptor substituents at P(III) and electron donor groups in aldehyde fragment stabilise the intermediates 3. The primary intermediates 3 containing chlorine atom at P(III) directly react with trialkyl orthoformates, acetals forming the P(IV) compounds, all being stable at room temperature. At the same time, intermediates without chlorine atom at P(III) decompose into 2 and 1 and the latter reacts with the nucleophiles. The adducts 3 were oxidised into P(IV) derivatives by using dimethyl sulphoxide and tert-butyl hypochlorite as oxidizers. We postulated that there might be a minimal concentration of 5 when its ability to protect the electrophilic centre of the carbonyl group is weakened, even though the catalysis is still present. Indeed, when the ratio 1, 2 and 5 is 1: 1 (or 2): 0.008 then the secondary intermediates $Cl_2POCHROCHRCl$ (R=Me, i-Pr, n-Pr) 4a and $(CCl_3CH_2O)_2POCHPr$-iOCHPr-iCl 4b are formed. At -20 °C 4 are stable. At temperatures higher that -10 °C due to low nucleophilicity of P(III) 4a easily decomposes into 1-chloro (1-chloroalkoxy)alkanes 6, which are separable by distillation. The intermediates 4a (R=Me, i-Pr, n-Pr) react with chlorine to produce 6 and $POCl_3$. Secondary intermediate 4b possesses enough nucleophilicity and undergo intermolecular Michaelis-Arbuzov isomerization: $(CCl_3CH_2O)_2POCHPr$-iOCHPr-iCl → $(CCl_3CH_2O)_2P(O)CHPr$-iOCHPr-iCl.

Chapter 2 – The review covers the reactions of aldehydes with monoterpenoids leading to chiral oxygen-containing heterocyclic compounds of various structural types. Almost all these reactions are catalyzed by acids, the Lewis acids and montmorillonite clays being typically used as the catalysts. Some of the resulting compounds exhibited significant biological activity. The most complex multistage transformations are observed when performing the reactions of aldehydes with oxygen-containing *para*-menthane, pinane and carane monoterpenoids in the presence of clays. Due to the ability of monoterpenoids of different structural types to rearrange into carbocations with the *para*-menthane framework upon protonation, a number of identical heterocyclic compounds can be obtained using various starting terpenoids.

Chapter 3 – This review provides information on the procedures and methodologies for the isolation, subsequent description and quantitative determination of aliphatic aldehydes, including the short- (C3-al to C10-al) and medium- (C12-al to C18-al) chain aldehydes and the long-chain ones (C22-al to C32-al). Aldehydes occur in numerous matrices, having both biological and commercial importance. Their physiological roles, the relevance of their presence and the incidence of their chain length are shown. Their relationship with other minor compounds such as alcohols and hydrocarbons is also reported. Special emphasis is put on the studies on the content and composition of the fractions containing the aldehydes in seed lipids and lipids from fruits and leaves. Also their presence in edible materials is commented. The different methodologies developed showed hexanal (C6-al) and nonanal (C9-al) as the most active compounds among the short-chain aldehydes, and hexacosanal (C26-al), octacosanal (C28-al) and dotriacontanal (C32-al) among the long-chain ones. In edible oils such as extra virgin olive oils, long-chain aliphatic aldehydes with even carbon-atom numbers from C22 to C30 are also present and can be isolated from the waxy fraction in quantities around a hundred mg kg^{-1} oil, being the C26 aldehyde the most abundant one in all of the studied samples.

Chapter 4 – In our effort to fermentatively produce biorenewable fuels and chemicals that are economically competitive with petroleum, it is desirable to use sugars derived from lignocellulosic biomass. However, the recalcitrance of this biomass requires some sort of depolymerization treatment to release the fermentable substrates. This depolymerization can be performed by enzymes, ionic liquids, hydrolysis or pyrolysis. Each depolymerization method has its own benefits and challenges. The challenge that is described here is the fact that hydrolysis and pyrolysis both result in the production of compounds that are inhibitory to the biocatalyst, limiting utilization of these carbon- and energy-rich streams. Specifically, we consider biocatalyst inhibition by aldehydes such as furfural, 5-hydroxymethylfurfural (5-

HMF), glycolaldehyde, syringaldehyde, vanillin and hydroxybenzaldehyde. Technologies exist for selectively removing these inhibitors, such as overliming or treatment with laccase enzymes, but these potentially increase the process cost. Selection-based strain improvement of a variety of industrially-relevant biocatalysts, including yeast and *Escherichia coli*, has provided insight into methods for strain improvement. Transcriptomic studies have provided complementary insights. Here, we describe these efforts and summarize both what is known regarding the mechanism of growth inhibition by these aldehydes and how biocatalysts can be engineered for increased tolerance.

Chapter 5 – Aldehyde autoxidation rates are very high even at room temperature and atmospheric pressure and this property has been often used to promote the so-called "co-oxidation" processes, in which a mixture of an aldehyde and another less reactive organic substrate is submitted to molecular oxygen. The present work is addressed to overcome the most recent developments in this field, covering the last decade. After a brief overview on significative co-oxidative processes mediated by aldehydes, we focus on the recent results reported by the authors for the selective epoxidation of olefins and the selective oxidation of alkylaromatic to the corresponding hydroperoxides promoted by the new *N*-hydroxyphthalimide (NHPI)/aldehyde catalytic system under mild aerobic oxidations. Enthalpic, polar, solvent and temperature effects are widely disclosed. Moreover, the free-radical reaction mechanism based on the in situ generation of phthalimido-*N*-oxyl (PINO) radical by means of the "molecule-induced homolysis" of NHPI is deeply investigated and supported by experimental evidences.

Chapter 6 – The reaction of gossypol with 2-amino-4,6-dioxypyrimidine in acidic environment has been studied. The structure of obtained product has been investigated by UV-, IR- and ^{1}H NMR-spectroscopy methods. Obtained results suggest that gossypol with 2-amino-4,6-dioxypyrimidine in acidic environment forms not C-N but C-C product due to keto-enol and lactam-lactim tautomerism of the latter.

In: New Developments in Aldehydes Research
Editors: L. Torrioni and E. Pescasseroli
ISBN: 978-1-62417-090-4

Chapter 1

SYNTHESIS AND PROPERTIES OF INTERMEDIATES IN REACTIONS OF ALDEHYDES WITH P (III) CHLORIDES

M. B. Gazizov, R. A. Khairullin, R. F. Karimova, and K. S. Gazizova

Organic Chemistry Department,
Kazan National Research Technological University, Kazan

ABSTRACT

For over a century the reaction of phosphorus trichloride 1a with aldehydes 2, which was discovered by Fossek, has been known and studied. An aqueous work-up or heating of the reaction mixture provides organic compounds with a P(IV) - C bond. All attempts had failed to determine the structure of intermediate products formed at the initial stages of this intricate reaction. Page, Kabachnik and Atherton postulated phosphite structure $[(RCHClO)_nPCl_{3-n}$, n = 1, 2, 3] 3 as primary intermediates, but it was thought that no spectrometric or chemical evidence for the intermediacy of 1 could be found due to their easy addition of second aldehyde molecule and transformation into the secondary intermediate $X_2POCHROCHRCl$ 4. Phosphorus (III) chlorides 1 were found to contain HCl as a contaminant, even after multiple distillations. HCl can be removed from 1 by treating with tertiary amine 5 or alkyl vinyl ether (AVE). The synthesis of 3 and a secondary possible intermediate 4 using low temperatures and purified 1is reported. Purified 1 does not react with 2. Under the same conditions commercial 1 reacts with 2 extremely exothermically showing the HCl catalysis. In turn AVE or a small amount of a tertiary amine gives rise to a violent reaction of 2 with purified 1 showing base catalysis. It was found that the addition of one or two molecules of 2 to one P - Cl bond depends on the concentration of 5. When 1-5 % of 5 is used, the adducts3 drived from the addition of one molecule of 2 are formed. The intermediate 3 reverts to 1 and 2 on heating, thus the reactions are reversible. We suppose that 5 not only makes the carbonyl oxygen more reactive but also protects its electrophilic centre from the addition of a second aldehyde molecule. Ten primary intermediates 3 were synthesized and their structures were confirmed by ^{1}H, ^{13}C and ^{31}P NMR. It has been shown that electron acceptor substituents at P(III) and electron donor groups in aldehyde fragment stabilise the intermediates 3.

The primary intermediates 3 containing chlorine atom at P(III) directly react with trialkyl orthoformates, acetals forming the P(IV) compounds, all being stable at room temperature. At the same time, intermediates without chlorine atom at P(III) decompose into 2 and 1 and the latter reacts with the nucleophiles. The adducts 3 were oxidised into P(IV) derivatives by using dimethyl sulphoxide and tert-butyl hypochlorite as oxidizers.

We postulated that there might be a minimal concentration of 5 when its ability to protect the electrophilic centre of the carbonyl group is weakened, even though the catalysis is still present. Indeed, when the ratio 1, 2 and 5 is 1: 1 (or 2): 0.008 then the secondary intermediates $Cl_2POCHROCHRCl$ (R=Me, i-Pr, n-Pr) 4a and $(CCl_3CH_2O)_2POCHPr$-iOCHPr-iCl 4b are formed. At -20 °C 4 are stable. At temperatures higher that -10 °C due to low nucleophilicity of P(III) 4a easily decomposes into 1-chloro (1-chloroalkoxy)alkanes 6, which are separable by distillation. The intermediates 4a (R=Me, i-Pr, n-Pr) react with chlorine to produce 6 and $POCl_3$. Secondary intermediate 4b possesses enough nucleophilicity and undergo intermolecular Michaelis-Arbuzov isomerization: $(CCl_3CH_2O)_2POCHPr$-iOCHPr-iCl → $(CCl_3CH_2O)_2P(O)CHPr$-iOCHPr-iCl.

Introduction

The reactions of P(III) chlorides with various carbonyl compounds in presence and absence of the third reagent are in detail discussed in review [1] and the monography [2] is devoted to the reactions of P(III) chlorides with O-containing organic reagents [2]. In the latter consideration of the interaction of carbonyl compounds with chlorides P(III) makes one of the basic sections. More than 20 years have already past since they had been published. For this time we had performed additional researches of reactions of chlorides P(III) with aldehydes, and new important data about key intermediates in the mechanism of their interaction were obtained. Therefore this chapter is devoted to the synthesis of primary and secondary intermediates and to the study of some of their transformations.

The chapter consists of five sections:

1) The Brief Historical Survey of Reactions of Aldehydes with P (III) Chlorides;
2) Synthetic Methods and Structure of Primary Intermediates;
3) Chemical Properties of the Primary Intermediates;
4) Reactions of the Primary Intermediates with Aldehydes and Detection of the Secondary Intermediates
5) Description of Experiments.

1. The Brief Historical Survey of Reactions of Aldehydes with P(III) Chlorides

In 1884 Fossek has found, that phosphorus trichloride1 vigorously interacts with aldehydes 2 with formation of dense oily addition products [1-3] which were called an adduct due to the uncertainty of their structure.

$$>PCl + RCHO \rightleftharpoons [adduct]$$

Later phosphorus tribromide 4 and triodide 5, phosphorochloridites 5, chlorophosphines 6 and other P(III) halides were also used.

Michaelis has attributed to adduct a structure of cyclophosphorane 7 [4]. Conant at first completely adhered to the Michaelis scheme [5]. However on the basis of kinetic researches he made an assumption about initial formation of a product of bipolar structure 8 [6], capable to transform into phosphorane 7.

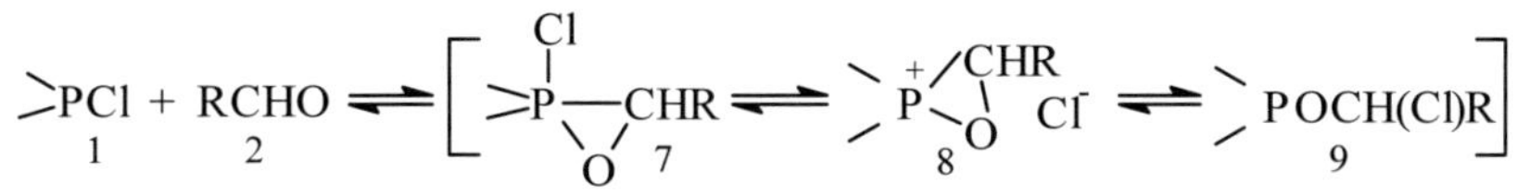

Page [7], Kabachnik [8-10] and Atherton with colleagues [11] held opposite point of view [11]. They considered, that chloride P(III) is electrophile and coordination number of phosphorus atom preserves in primary process. Reaction products are P(III) I-chloroalkoxyl derivatives 9.

In the case of phosphorus trichloride 1 a gradual substitution of chlorine atoms by1-chloralkoxyl group and formation of tri(1-chloroalkyl) phosphites 12, which then isomerise in P(IV) derivatives 13 [8-10], were supposed.

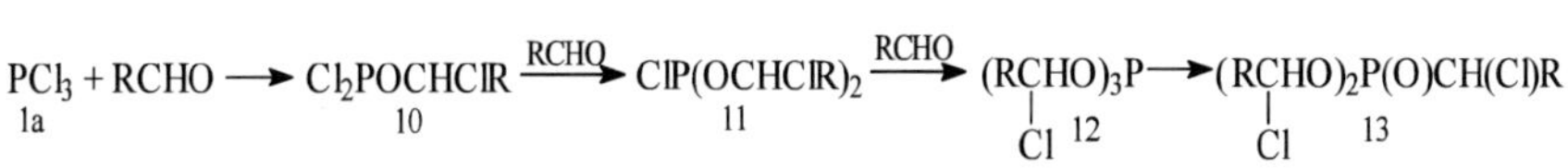

Pudovik and Khairullin held the opinion of nucleophilic character of P(III) chlorides in the primary act of interaction [12]. However, taking into account the data of works [7-11] they have complemented the Michaelis-Conant scheme with a product 9.

If to subject the adduct to pyrolysis, the stable P(IV) 1-chloroalkyl derivatives 14 [8-10] are formed. Hydrolysis of the adduct in soft conditions leads to the formation of 1-hydroxyalkanephosphonic acids 15 [5,7,11-19].

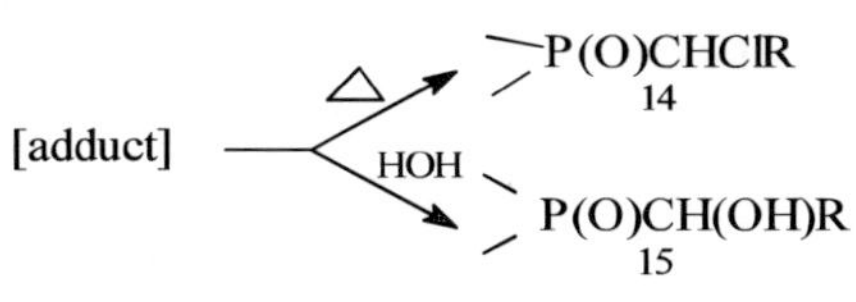

On the basis of a structure of I-chloroalkyl- and I-hydroxyalkyl derivatives of P(IV) 14-15 the researchers assumed, that at the adduct formation only one molecule of aldehyde is expended for each P-C1 bond. However, since 70th years of the last century the publications have appeared confirming the expenditure of two molecules of aldehyde per one P-Cl bond [20-26].

On the basis of influence of substituent nature at P(III) in compounds 1 on the ease of reaction proceeding the conclusion has been made that PCl_3 and P(III) dichlorides show electrophilic reactivity [20-23], bat monochlorophosphines show nucleophilic reactivity [23-25] . Phosphorus tribromide and triiodide are more active electrophiles than PCl_3 [23].

As to phosphorochloridites, their reactivity depends on the nature of hydrocarbon radicals bound with oxygen atoms. If a radical is phenyl (diphenylphosphorochloridite) or phenylene-1,2 (catecholchlorophosphite), P(III) monochloride shows electrophilic reactivity [26]. If they are alkyl radicals, dialkyl phosphorochloridites show both nucleophilic and electrophilic reactivity at differet stages of reaction [27].

It is necessary to notice, that such order of dependence of P(III) chlorides reactivity is also observed in towards anhydrides and monoacylals of carboxylic acids. Tri- and dichlorides, as well as catecholchlorophosphite, dialkyl- and diarylphosphorochloridites show electrophilic reactivity. At the initial stage of reaction chlorine atom at P(III) is substituted by acyloxy-group [28-34]. Diethylchlorophosphine shows nucleophilic reactivity towards the above-mentioned derivatives of carboxylic acids [34-35].

Detailed research of the reactionsof $PHal_3$ (Hal = Cl, Br, I) with ethanal and aromatic aldehydes 4-X-C_6H_4CHO (X = H, MeO, NO_2) by NMR method was performed by J.A Miller with coworkers [20 - 21, 23]. They have established, that electrophilic reactivity of P(III) halides increases by replacement of chlorine by bromine and bromine by iodine. In all cases at the beginning the formation of di-α-chloroether 16 proceeds which stability depends on the nature of radical R. If R are aliphatic ones, they are stable. Compounds 16 with an aromatic radical, especially at heating (70 °C) transform into arylidene dihalides 17. Their stability sharply falls by replacement of chlorine by bromine and iodine.

$$\underset{1,3,4}{PHal_3} + \underset{2}{RCHO} \longrightarrow \underset{16}{(RCHHal)_2O} + \underset{18}{(POHal)_n}$$

$$16 \longrightarrow RCHO + \underset{17}{RCHHal_2}$$

$$17 + 18 \xrightarrow{\triangle} \underset{14}{Hal_2P(O)CHHalR}$$

16, 17, R=Ph (a), 4-$NO_2C_6H_4$ (b), 4-$MeOC_6H_4$ (c), Me (d).

Di(α-halogenobenzyl ether 16 decomposes into arylidene dihalide 17 and starting aldehyde 2 which reacts again with $PHal_3$. It is proposed, that $PHal_3$ transforms into a polymeric product $(POHal)_n$ 18 possessing sufficient nucleophilicity to react with arylidene dihalide 17 at high temperature (150-180 °C) in accordance with Michaelis-Arbuzov reaction with the formation of a product 14. However, the question of the structure and properties of phosphorus-containing product 18a is still open, though there are data on its synthesis by independent way from PCl_3 and acetic anhydride and on its reaction with α-chlorobenzylacetate in mild conditions [24]. It is well-known that [36-37] only P(III) derivatives enter into the Michaelis-Arbuzov reaction. The ^{31}P NMR spectra data of reaction mixtures ofinteraction of PCl_3 with benzaldehyde [21] and aliphatic aldehydes [22] contain a set of resonace signals at 38-48 ppm corresponding more to phosphorus atoms in phosphoryl compounds with P-C bond, but not to threecoordinated phosphorus atom. It is also noted that $PhCHBr_2$ + $(POBr)_n$ system is less inclined to the formation of $Hal_2P(O)CHHalPh$, than that of $PhCHCl_2$ + $(POCl)_n$ system [23], though it is known that benzylidene dibromide is an more active electrophile than benzylidene chloride. Finally, we have found out that at 120-130 °C arylidene dihalides ($ArCHHal_2$ Hal=Cl, Br) do not interact with trialkyl phosphites. Phosphite 19 isomerises into phosphonate 20, and *gem*-dihalide 17 remains unchanged [38-39]. Moreover, at higher temperatures (160-180 °C) the phosphonate 20 formed reacts with compound 17 producing aldehyde 2 and cyclic trimer of alkylphosphonic acid anhydride 21. Reactions between compounds 17 and 19 at 160-180 °C lead to the same products 2 and 21.

$$\underset{17}{ArCHHal_2} + \underset{19}{(RO)_3P} \xrightarrow{120\text{-}130^{o}} \underset{20}{RP(O)(OR)_2} + 17$$

$$RP(O)(OR)_2 + ArCHHal_2 \xrightarrow{160\text{-}180^{o}} \underset{2}{ArCHO} + \underset{21}{(RPO_2)_3} + RHal$$

$$ArCHHal_2 + (RO)_3P \xrightarrow{160\text{-}180^{o}} \underset{2}{ArCHO} + \underset{21}{(RPO_2)_3} + RHal$$

Nurtdinov S.Kh. with coworkers studied interaction of benzaldehyde with PCl_3 (2: 1, reflux , 4 hrs), ethyldichlorophosphine 22 (2:1, 6 hrs) and diphenylphosphorochloridite 23 (1: 1, room temperature, 120 hrs) [40-41]. From the reaction of chloride 23 diphenyl ether of α-(α-chlorobenzyloxy)benzylphosphonic acid 24 was isolated. In the case of ethyldichloro-phosphine the adduct was treated by phosphorus pentachloride in benzene and ethyl-α-chlorbenzylphosphinic acid chloroanhydride 14a was obtained by distillation. They synthesized earlier known acid 15a [42] by treating with water the hygroscopic crystal product. The crystals were precipitated from reaction mixture of benzaldehyde interaction with PCl_3 in pentane.

$$\underset{23}{(PhO)_2PCl} + 2\,PhCHO \longrightarrow \underset{24a}{(PhO)_2P(O)CH(OCHClPh)Ph}$$

$$\underset{22}{EtPCl_2} + 2\,PhCHO \longrightarrow [adduct] \xrightarrow[-\,POCl_3]{PCl_5,\ C_6H_6} \underset{14a}{Et(Cl)P(O)CHClPh}$$

$$PCl_3 + 2\,PhCHO \longrightarrow [adduct] \xrightarrow{H_2O} \underset{15a}{(HO)_2P(O)CH(OH)Ph}$$

It follows from the works [40-41] that, firstly, the formation of phosphoryl compounds with P-C bond in reactions of benzaldehyde with P(III) chlorides occurs at much lower temperatures (25-80 °C) in comparison with earlier applied ones (150-180 °C) [9 - 10, 20 - 21]; secondly, results are interpreted without involving phosphorus-containing polymer $(POCl)_n$ as intermediate with nucleophilic properties.

The use of ^{1}H and ^{31}P NMR method gave new results in the study of the interaction of P(III) chlorides with saturated aliphatic aldehydes (ethanal and butanal, in 1: 2 ratio) [22, 40].

The reactions were performed in soft conditions: in benzene at 20 °C. In ^{31}P NMR spectra of reaction mixtures the resonance signals at 14, 38, 48 ppm (in the case of PCl_3) and at 38 and 48 ppm (in the case of $PhPCl_2$) were observed.

These resonance signals correspond to phosphorus atoms in phosphoryl compounds with one and two P - C bonds. To transform the unstable intermediate compounds >P(O)OCHROCHClR into more stable ones, the adduct was treated with phosphorus pentachloride for the first time.Phosphonic acid dichloroanhydrides 25 and the products of their dihydrochlorination 26 were synthesized in the case of PCl_3.

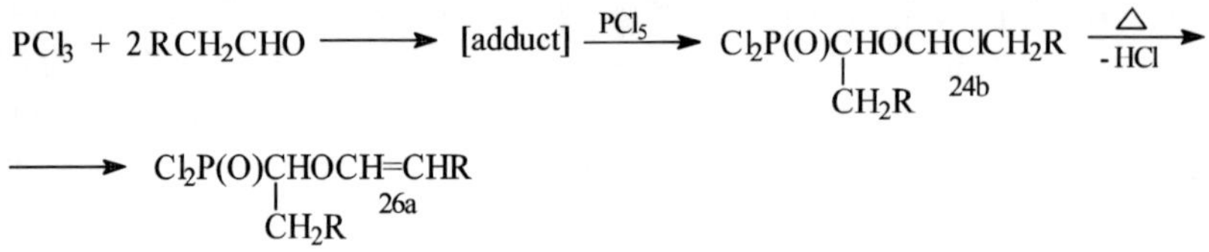

In the case of reaction phenyldichlorophosphine 27 with butanal the following mechanism [22] has been proposed:

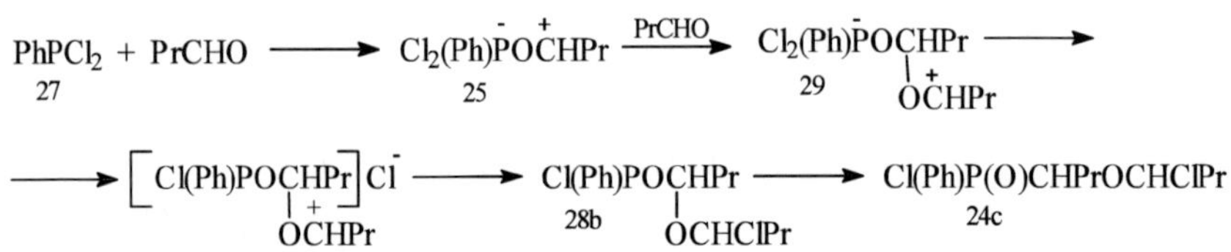

The absence of primary intermediate Cl(Ph)POCHClPr structure, similar to that occuring in the case of PCl_3 ($Cl_2POCHClR$) is one of the drawbacks of this scheme [7-9, 11]. And the advantage of the scheme is that it contains for the first time the structures of possible secondary intermediates 28b and 28c. Similarly to the structures 28b and 28c we have adopted the structure of secondary intermediate $Cl_2POCHROCHClR$ 28a [43 - 44] for the reaction of PCl_3 with aliphatic aldehydes.

Based on the foregoing for the reactions of P(III) chlorides showing electrophilic properties with aldehydes in mild conditions (25-80 °C) the following general scheme can be proposed:

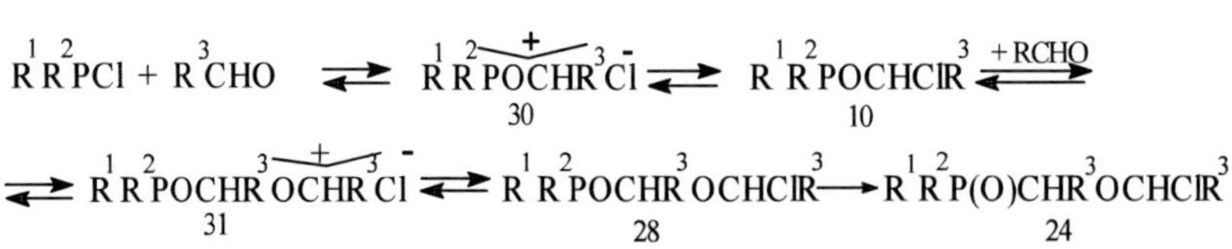

The product of addition of P(III) chloride to one molecule of aldehyde or replacement of chlorine atom at P(III) by 1-chloroalkoxyl group (primary intermediates 10, 11or12) is formed as a result of nucleophilic attack of carbonyl group oxygen atom on electrophilic P(III). That leads to the formation of ionic pair 30. The last one adds the next molecule of aldehyde forming new ionic pair 31 with two fragments of aldehyde, and then the compound 28 containing 1-(chloroalkoxy)alkoxyl group at P(III). The product 28 represents secondary intermediate with P(III). When its nucleophilicity is high enough, it isomerizes into P(IV) derivative 24.

Thus, the formation of phosphoryl compounds with P - C bond 24, which have been earlier trasformed into the 14 and 15 by corresponding treatment, occurs already at 25-80 °C . As regards a synthesis of compounds 14 at 150-180 °C, firstly, their yileds are very low, secondly, they can be formed at high-temperature decomposition of compounds 24. The question on a structure of a polymeric product $(POHal)_n$and its participation in the reaction with arylidene dihalides producing $Hal_2P(O)CHHalAr$ 14 remains unclear.

It is assumed that at different stages in reactions with aldehydes dialkylphosphorochloridites show both electrophilic and nucleophilic properties. Interaction results in the formation of polymeric products [27].

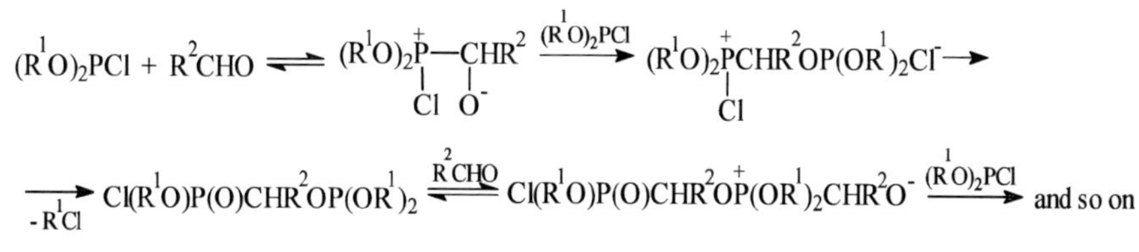

Quasiphosphonic salts 35 are obtained by interection of benzaldehyde with secondary chlorophosphines – P(III) chlorides showing exceptionally nucleophilic properties. These salts transform into P(IV) derivatives containing two aldehyde fragments 24d-e [23-25] in accordance with the second stage of Michaelis-Arbuzov reaction. At interaction of diphenylchlorophosphine with aliphatic polyfluoroaldehydes stable phosphoranes 34a [25] are formed.

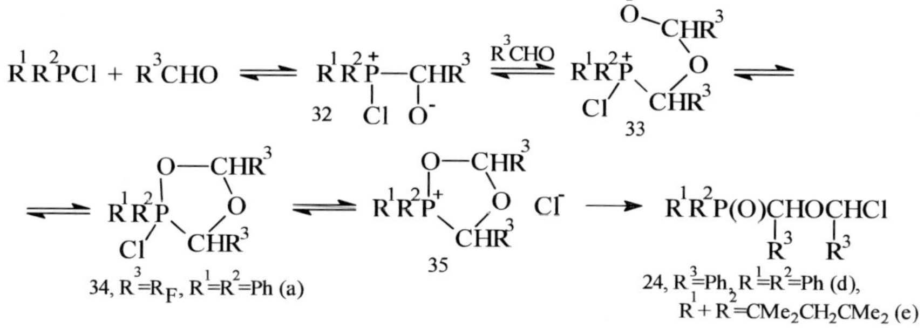

Phosphoranes 34b formed at interaction of dipropylphosphorochloridite with polyfluoroaldehydes stabilize by elimination of propyl chloride preserving 1,4,2-dioxaphospholane cycle (compound 36) [25].

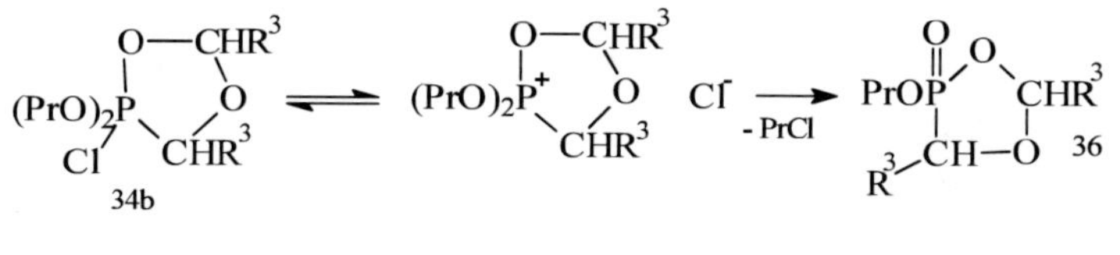

The formation of 34 and 35 can be presented as follows: as a result of P(III) lone pair nucleophilic attack on the carbonyl group carbon atom the bipolar ion 32 is formed. The ion 32 adds the second aldehyde molecule and turns into a new bipolar ion 33, which cycles into phosphorane 34.

Diethyl-1-(buthene-1-yloxy)buthylphosphine oxide 26b [22] in high yield was obtained by distillation of a product of interaction of diethylchlorophosphine with butanal. Obviously, as well as in the case of other secondary chlorophosphines at the beginning the product 24f is formed. Under heating *in vacuo* compound 24feliminates HCl and transforms into phosphine oxide 26b.

$$Et_2PCl + 2\,PrCHO \longrightarrow \underset{24f}{Et_2P(O)CHPrOCH(Cl)Pr} \xrightarrow[-\,HCl]{\Delta} \underset{26b}{Et_2P(O)CHPrOCH{=}CHEt}$$

Thus, for the reactions of aldehydes with P(III) chlorides showing electrophilic reactivity the primary and secondary intermediates with the fragments >POCHClR and >POCHROCHClR were predicted. But their synthesis has not succeeded until now.

2. Synthetic Methods and Structure of Primary Intermediates

Though the structure of primary intermediates 10 was considered conventional [7-12], they were found impossible not only to be isolated but also to be detected by spectral methods. They have come to an opinion that in the presence of acid admixtures the second aldehyde molecule adds very rapidly to intermediates, what makes the synthesis of intermediates 10 almost impossible. So, it was necessary to develop absolutely new approach to study these reactions aimed at separating the stages of addition of the first and second molecules of aldehyde to P-C1 bond.

It is necessary to emphasize, that catalytic effect of HCl admixture in P(III) chlorides has been found out in reactions with many O-containing organic compounds, which are alkene oxides [45-47], carboxylic acids [48], acetals [49-50] and aldehydes [51-52]. The P(III) chloride purified from HCl admixture by treating them with alkyl vinyl ethers (AVE) was found not to react with aldehydes and acetals at 20 °C. Reaction initiates only by bringing HCl in a reaction mixture in various ways [51-52]. Thus, only removal of acid impurity from P(III) chlorides does not make it possible to solve a problem of separation of addition stages of the first and second molecules of aldehyde to one P-Cl bond.

We knew that P(III) chlorides in particular catecholchlorophosphite 37 react with the aldehydes activated by such nucleophiles as trialkylphosphites 19 and triethylamine 38a expending one molecule of compound 2 for any P-Cl bond. New bond C-X (X = P [53-54], N [55-56] is formed. The coordination number of the phosphorus atom originally bound with chlorine does not change (see 39 and 40).

$$\underset{39\quad n=1,2,3}{[(R^1O)_2P(O)CHR^2O]_nPCl_{3-n}}$$

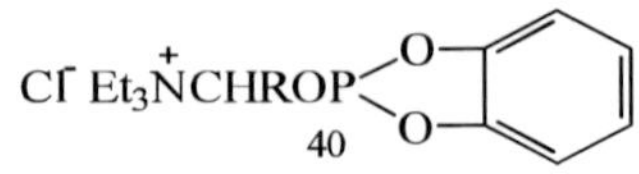

40

We have proposed that nucleophils 19 and 38a do not only activate carbonyl group making oxygen more nucleophilic, but also protect elecrophilic centre from addition of the second aldehyde molecule. We have also found [56] that reaction with the use of triethylamine is reversible: in soft conditions (-10 °C) salt 40 is stable enough and in progress of formation it precipitates. At 20 °C its decomposition becomes apparent, especially at continuous removal of decomposition products from the reaction mixture. So, if salt 40 is in contact with excess of triethyl phosphite for 30 days diphosphorus-containing compound 41 is formed according to the Birum reaction.

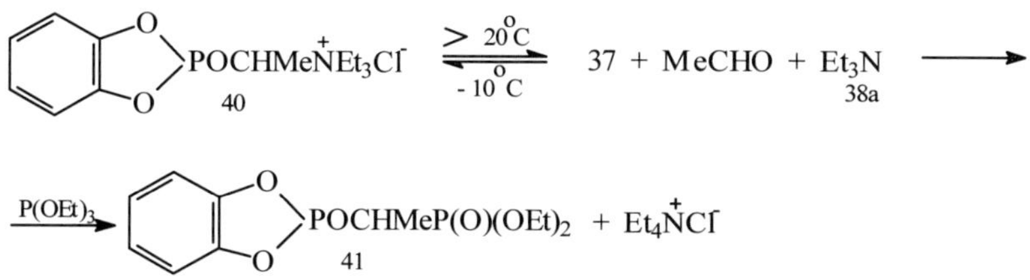

At heating (110-120 °C) in vacuum salt 40 completely decomposes to initial compounds and P(III) chloride 37 is obtained in a good yield. Knowing that the reaction is reversible when triethylamine is used, we assumed that using relatively small concentration of amines 38 should allow to synthesize primary intermediates.

Thus, the essence of the developed approach consists in removal of acid admixtures from P(III) chloride and in catalytic blocking of the electrophilic centre of carbonyl group.

The aim was achieved by the use of compounds showing basic and nucleophilic properties. Tertiary amines 38 and alkyl vinyl ethers 42 were used as such compounds.

At treating P(III) chloride with tertiary amine HC1 admixture turns into amine salt 43. Usually compounds 1 and 38 is mixed up at a room temperature in ratio 1.0: 0.10 - 0.15 and P(III) chloride is distilled directly into a reaction flask [43 - 44]. Amine salt 43 remains in a distillation flask.

$$>PCl \cdot HCl + \ddot{B}\ (38) \longrightarrow\ >PCl + B \cdot HCl\ (43)$$

$$>PCl \cdot HCl + CH_2{=}CHOR\ (42) \longrightarrow\ >PCl + CH_3CH(OR)Cl\ (44)$$

By other technique P(III) chloride and AVE mix up in the ratio 1:1 at 10°C. The α-chloroether 44 formed and AVE excess are removed *in vacuo* and compound 1 is distilled into a reaction flask [57]. AVE has appeared to be more suitable for refining the high-boiling P(III) chlorides but they can not be used for purification of PCl_3 since the latter vigorously reacts with AVE to give organophosphorus compound with P - C bond [57-58]. PCl_3 also causes polymerization of AVE. Tertiary amines have appeared to be more universal and they can be used for purifying practically all P(III) chlorides including PCl_3.

And it is possible to use low-boiling (trialkylamines) or high-boiling (N, N-dialkylanilines) organic bases depending on boiling temperature of P(III) chloride.

We synthesized the primary intermediates which have been divided into three types according to P(III) environment:

$RCHClOPCl_2$

10, R=Me(a), n-Pr(b), i-Pr(c), CCl_3(d)

$R^1CHClOP(OR^2)Cl$

11, R^1=Me, R^2=MeCHCl(a)

R^1= i-Pr, R^2=CCl_3CH_2(b)

$R^1CHClOP(OR^2)(OR^3)$

12, R^1=Me,R^2=R^3=MeCHCl (a)

R^1= i-Pr, R^2=R^3=CCl_3CH_2 (b), Ph(c), R^2=CCl_3CH_2, R^3= i-PrCHCl (d)

R^1=Me, R^2=CCl_3CH_2, R^3=MeCHCl (e)

R^1= i-Pr, R^2=R^3=Me (k), Et (l)

R^2=R^3= 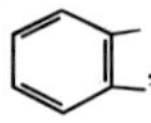, R^1=Me (g), i-Pr (f), n-Pr (h), t-Bu (i).

Intermediates with P-C bond have appeared rather unstable, and we have failed to synthesize them so far.

In order to synthesize intermediate 10a, PCl_3 purified was brought in reaction with aldehydes. However the formation of phosphorus-containing products was not observed. Phosphorus trichloride caused only aldehyde trimerization. In ^{31}P NMR spectrum of reaction mixture only the resonance signal at δp 220 ppm was observed corresponding to phosphorus atom in PCl_3. In ^{1}H and ^{13}C NMR spectra the resonance signals corresponding to hydrogen and carbon nuclei in aldehyde trimers $(RCHO)_3$ 45 were only observed. For example, in the case of acetaldehyde paraldehyde is formed and it was isolated from reaction mixture in a good yield.

The result of interaction between compounds 1 and 2 in presence of catalytic amounts of tertiary amine drastically depended on the order of mixing of the reagents. At addition of 3-4 drops of PCl_3 to a mixture of aldehyde and amine violent polymerization of compound 2 proceeded and the formation of phosphorus-containing organic product did not occur. Addition of the rest amount of PCl_3 did not result in the temperature rise. After removal of PCl_3 in vacuum paraldehyde was obtained in a good yiled.

A slow addition of aldehyde to the mixture of PCl_3 with a small amount of tertiary amine (triethylamine, N, N-dimethyl- and -diethylanilines, pyridine, 2,6-dimethylpyridine) gave rise to violent reaction showing base catalysis. In 2 hrs after mixing the reagents in a ^{31}P NMR spectrum of a reaction mixture three resonance signals were observed: δp220 (PCl_3) and 177.1, 177.8 ppm (3:1). The last two ones have very close chemical shifts and correspond to a phosphorus environment ($OPCl_2$) in alkylphosphorodichloridites [59-60]. We ascribed both last signals to P(III) in intermediate 10a [43-44].

$$\underset{1}{PCl_3} + n\,\underset{2}{RCHO} \rightleftharpoons \underset{10\text{-}12}{(RCHO)_nPCl_{3-n}}$$

10, n=1, R=Me (a), n-Pr (b), i-Pr (c);

11, n=2, R=Me (a); 12, n=3, R=Me (a).

In 6 hrs after mixing of the reagents ratio PCl_3: 10a makes up 1.4:1.0 from the integral intensity of these signals and further it doesn't practically change. According to ^{1}H NMR after 5 hrs a ratio of the substances 2a, 4a and 10a makes up 1:9:40.

After removal of PCl_3 and insignificant admixture of acetaldehyde the residne was subjected to distillation *in vacuo* using Arbuzov flask or flask equipped with Vigreux column of 10-15 cm length. The mixture of compounds 10a and 45a (5:1) was obtained. ^{31}P NMR (δp, ppm): 178.1 and 177.8, ^{13}C NMR (δ, ppm): 19.89 s, CH_3, 97.5 s, CH, 45a; 24.88 s, CH_3, 86.06 d, $^2J_{PC}$ 10.6 Hz, CH, 10a.

For successful synthesis of compounds 10a-c the best ratio of initial reagents PCl_3, aldehyde and tertiary amine is 1:1:0.03-0.05. Aldehyde is slowly added dropwise to intensively stirred mixture of purified PCl_3 and amine at -5 - 0 °C. After removal of the volatile products *in vacuo* distillation of crude product afforded pure 10в-c. They are also stable at a room temperature and can be stored in a sealed ampoule for several months. The structure of intermediates 10b-c was confirmed by their complete elemental analysis, ^{1}H, ^{13}C and ^{31}P NMR spectra (Table 2.1) and chemical transformations (Section 3).

When the amount of aldehyde is increased the substitution of chlorine atoms at P(III) turns out to be non-selective [43-44]. Thus, when the ratio of 1:2:38 is 1:2:0.01 the composition of the reaction mixture (ascertained by ^{31}P NMR spectroscopy) is 5.4% 1, 39.9% 10a and 55.3% 11a. ^{31}P NMR (δp, ppm): 218 (PCl_3); 178.1 , 177.8 (3:1 , 10a); 163.30 ,162.95 ,162.43 (3:2:1).The third set of resonance signals corresponds to a phosphorus environment O_2PCl in dialkyl phosphorochloridites [60].

Therefore these three signals were ascribed to P(III) in intermediate 11a. After cautious removal of volatile products by suction in trap with liquid N_2 at room temperature crude 11a was obtained. ^{31}P NMR (δp, ppm): 163.3, 162.95 and 162.43; ^{1}H NMR (δ, ppm): 1.83 d, $^3J_{HH}$ 6.4 Hz, 6H, CH_3; 6.14 m , 2H , CH (it is shifted to the higher fields in comparison with methylene proton in 10, 6.38 ppm). ^{13}C NMR (δ, ppm): 27.66 s and 27.51 s, CH_3; 84.66 d, $^2J_{PC}$ 8.2 Hz, 84.56 d, $^2J_{PC}$ 5.3 Hz, CH.

The structure of intermediate 11a is confirmed by satisfactory elemental analysis and by their oxidation to the corresponding phosphate (Section 3). The mixture condensed in a liquid nitrogen trap according to the NMR spectra consists of PCl_3, trimer 45a and intermediate 10a.

While trying to distill the substance 11a*in vacuo* it gradually decomposes and almost nothing remains in distillation flask. According to ^{1}H and ^{31}P NMR spectra (recorded at -10 °C) the decomposition products condensed in a trap with liquid nitrogen are the compounds 1, 2a and 10a. Special experiment has demonstrated that at long-term heating *in vacuo* to 100 °C intermediate 10a also completely decomposes into PCl_3 and aldehyde.

At further increase of aldehyde amount the mixture of products of various substitution degree of chlorine atoms at P(III) by 1-chloroethoxyl group is formed [43-44]. In ^{31}P NMR spectrum of reaction mixture recorded in 24 hrs after mixing the reagents two new resonance signals at δp 146.9 and 138.7 ppm (1:5) appear in addition to those from phosphorus nuclei in compounds 10a and 11a. They correspond to the environment of phosphorus $O_3P(III)$ in trialkyl phosphites [59-60].

Therefore they were ascribed to P(III) in one more intermediate 12a i.e. in tri(1-chloroethyl) phosphite. When the ratio of 1:2:38 is 1:4:0.05 then the mixture contains 3.1% 10a, 24.8% 11a and 72% 12a.

Table 2.1. ^1H, ^{13}C and ^{31}P NMR data for primary intermediates R^1 R^2 $POCHClR^3$

N Comp-ds	R^1	R^2	R^3	^mE	δ, ppm ($^nJ_{HH}$ and $^nJ_{PH}$, Hz)					$δ_P$, ppm
					E-1	E-2	E-3	E-4	E-5	
1	**2**	**3**	**4**	**5**	**6**	**7**	**8**	**9**	**10**	**11**
10a	Cl	Cl	CH_3	^1H ^{13}C	-	6.38 dk (5.6, 13.9) 86.06 (d) (10.6)	1.81 d	-	-	178.1 177.7
10b	Cl	Cl	n-Pr	^1H	-	6.35 dt (6.5, 15.0)	2.1 dt (6.5, 7.5)	1.55 secstet (7.5, 7.5)	0,94 t (7.5)	180.42 180.83
10c	Cl	Cl	$CH(CH_3)_2$	^1H ^{13}C	-	6.13 dd (3.8, 6.7) 94.4 d (10.0)	2.2 dheptet (3.8,6.7) 36.87 s	1.06 d, 1.08 d (6.7) 17.27 s, 16.63 s	-	179.61 179.24
10d	Cl	Cl	CH_3	^1H	-	6.33 d (11.5)	-	-	-	184
11a	Cl	$OCHClCH_3$	CH_3	^1H ^{13}C	-	6.14 m 84.66 d, 84.56 (d) (8.2)	1.83 d (6.4) 27.66 s, 27.51 s	-	-	166.32 162.97 162.43
11b	Cl	CCl_3CH_2O	$HC(CH_3)_2$	^1H	4.1 d (5.0)	6.14 dd (4.0, 10.0)	2.1 dheptet (4.0,7.0)	0.9 d (7.0)	-	164.0 161.0
12a	$OCHClCH_3$	$OCHClCH_3$	CH_3	^1H ^{13}C	-	6.14 m 84.11-84.5 m	1.81 d (5.6) 27.56-27.77 (m)	-	-	145.85 138.82
12b	CCl_3CH_2O	CCl_3CH_2O	$HC(CH_3)_2$	^1H	4.1 d (5.0)	4.95 dd (5.0, 9.0)	2.1 m	0.95 d (6.3)	-	133.0
12c	PhO	PhO	$HC(CH_3)_2$	^1H	7.1 m	5.6 dd (4.5,9.0)	1.93d,heptet (4.5, 6.5)	0.9 d (6.5)	-	138.0 141.0

N Comp-ds	R^1	R^2	R^3	${}^{m}E$	δ, ppm (${}^{n}J_{HH}$ and ${}^{n}J_{PH}$, Hz)					δ_P, ppm
					E-1	E-2	E-3	E-4	E-5	
1	**2**	**3**	**4**	**5**	**6**	**7**	**8**	**9**	**10**	**11**
12d	CCl_3CH_2O	OCHCl $(CH_3)_2$	$HC(CH_3)_2$	${}^{1}H$	4.2 d (5.0)	6.15 m	2.10 m	1.0 d (6.5)	-	139.0 142.0
12e	CCl_3CH_2O	$OCHClCH_3$	CH_3	${}^{1}H$	4.1 d (5,0)	6.14 m	1.85 d (6.5)	-	-	138.0 141.0
12f	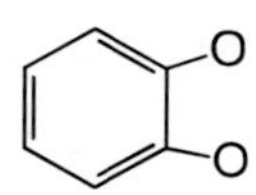		$CH(CH_3)_2$	${}^{1}H$	6.7 m	5.5 dd (4.5, 9.0)	1.93 dheptet (4.5,6.5)	0.9 d (6.5)	-	133.0
				${}^{13}C$	109.3 s 120.5 s 142.3 d (6.5)	88.5d (6.5)	33.5 (s)	13.5 s		
12g	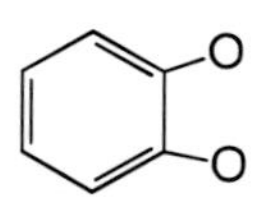		CH_3	${}^{1}H$	6.8 m	5.4 dk (6.3, 7.5)	1.37 d (6.3)	-	-	133.1
				${}^{13}C$	110.0 s, 120,5 s, 142.0 d (6.3)	80.8 d (7.5)	24.0 s			
12h	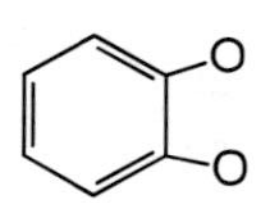		$C(CH_3)_2$	${}^{1}H$	6.8 (m)	5.3 d (9.0)	-	0.9 s	-	133.0

When trying to distill the mixture *in vacuo* it quite rapidly decomposes into 10a, PCl_3 and acetaldehyde. The structure of intermediate12a was confirmed by ^1H, ^{31}P, ^{13}C NMR spectra (Table 2.1) and by its transformation into chlorophosphate 65d (Section 3).

Here it is necessary to underline that the assumed earlier [8-10] isomerization of compound 12a into P(IV) derivative is not realized. Both 11a and 12a revert to PCl_3, 2 and 10a on heating, thus the reactions are reversible.

Thus, for the first time three types of primary intermediates 10a, 11a and 12a were synthesized by the interaction of PCl_3 with aldehydes in the presence of catalytic amount of tertiary amine. They differ from each other by substitution degree of chlorine atoms in PCl_3 by 1-chloroalkoxyl group. The formation of intermediates 10-12 is a reversible process. Accumulation of 1-chloroalkoxyl groups at P(III) reduces their stability: intermediates 11a and 12a exist only in the cold. At room temperature they decompose into initial compounds. In the presence of extra amount of aldehyde in the system these intermediates produce the P(IV) derivatives (Section 4). The intermediates 10a-care distilled *in vacuo*, stored for a long time in a sealed ampoule and decompose only at long-term heating *in vacuo*.

The second method for the synthesis of primary intermediates is associated with the application of AVE [57-58]. At the beginning catecholchlorophosphite (CCP) was used without purification from HCl admixture. It was mixed with AVE at 10 °C and 2-methylpropanal was added dropwise to this mixture.

But it turned out that in the ^1H NMR spectrum of even the distilled product 12j the multiplet resonance signals were found out at δ 1.0 and 3.9 ppm, which intensities changed from one experiment to another. These signals were ascribed to the AVE polymerization products.

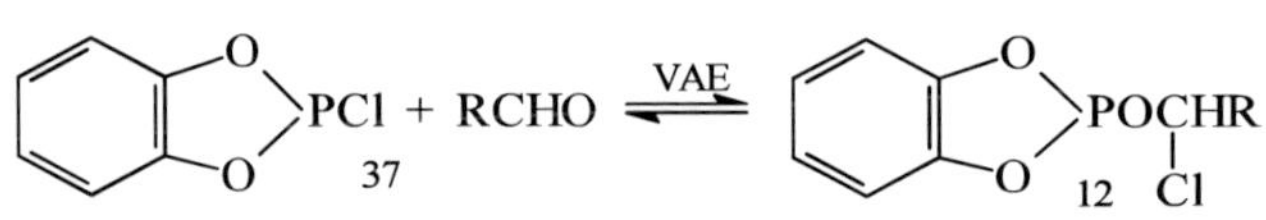

12, R= i-Pr (f), Me (g), n-Pr (h), t-Bu (i)

To remove these admixtures, CCP after additional purification with AVE was distilled into reaction flask, where AVE and aldehyde were placed. The compound 12f not containing after distillation other admixtures except 15 % of initial CCP was synthesized. The improved method was used for the synthesis of new intermediates 12g-i. The compound 12i was isolated by distillation *in vacuo* (Table 2.1). The others were investigated as crude products after removal of volatile substances *in vacuo* (Table 2.1).

Diphenyl phosphorochloridite 23 reacts with 2-methylpropanal in a similar way. According to ^{31}P NMR ratio of the compounds 12c (δp 130 ppm) and 23 (δp 156 ppm) made up 10:1.

$$\underset{23}{(PhO)_2PCl} + Me_2CHCHO \xrightleftharpoons{VAE} \underset{12c}{(PhO)_2POCH(Cl)CHMe_2}$$

To confirm the structure of intermediate 12c aldehyde was quickly removed in the cold *in vacuo* and NMR spectrum of the residue was recorded (Table 2.1).

We found out that conversion degree of initial products and stability of intermediates 12 depend on the aldehyde structure. When a ratio CCP: RCHO: AVE is 1:2:2 conversion of CCP made up 58 , 65 , 75% at R=Me, n-Pr, i-Pr, respectively. Intermediate 12i (R=t-Bu) was distilled without decomposition. The compounds 12 f-h decomposed by distillation and the CCP content in distillate made up 15, 40, 50% at R=i-Pr; n-Pr and Me, respectively. These data testify that stability of intermediates 12 increases with the increase of nucleophilic properties of aldehyde. The stability of primary intrermediates depends also on the nature of substituents at P(III). By replacement of two electron-acceptor chlorine atoms at P(III) in 10a by more electron-donor 1-chloroalkoxyl (in 11a and 12a), phenoxyl (in 12c) and 1,2-phenylenedioxyl (in 12f-h) groups the stability of these intermediates decreases.

It is known that tris(2,2,2-trichloroethyl) phosphite and phenylbis(2,2,2-trichloroethyl) phopshonite do not enter the Michaelis-Arbuzov reaction even with such strong electrophils as α-chloroethers [61]. Obviously, trichloroethyl group possesses rather high electron-acceptor ability and nucleophilicity of P(III) sharply decreases. We supposed that replacement of two 1-chloroethoxyl groups at P(III) in 12a by 2,2,2-trichloroethoxyl ones will increase the stability of primary intermediate 12b. It was synthesized by interaction bis(2,2,2-trichloroethyl) phosphorochloridite 46a with 2-methylpropanal in the presence of AVE as the catalyst. 12b was found to be stable at a room temperature for a long time.

$$\underset{\textbf{46a}}{(CCl_3CH_2O)_2PCl} + i\text{-}PrCHO \xrightleftharpoons{VAE} \underset{\textbf{12b}}{(CCl_3CH_2O)_2PO}\underset{\displaystyle Cl}{\underset{|}{C}}HPr\text{-}i$$

But at heating *in vacuo* it decomposes with elimination of aldehyde molecule even before reaching the boiling temperature. Therefore 12b was identified as crude product after removal of all volatile admixtures *in vacuo* (Table 2.1). It was also synthesized by the third indirect method: by interaction of intermediate 10c with 2,2,2-trichloroethanol in the presence of tertiary amine (Section 3). NMR spectral data for 12b obtained by different methods are identical. Attempts to synthesize primary intermediates from ethyl phosphorodichloridite and ethyldichlorophosphine using direct method have failed. The reaction mixture represented very viscous liquid and in its NMR spectra did not observe resonance signals corresponding to P(III) and protons of CH_3CH (C1)OP(III) fragment. Probably, primary intermediates containing fragments EtO[MeCH(Cl)O]P(III) and Et[MeCH(Cl)O]P(III), are even less stable than compounds 11a and 12a. There is another aspect of the problem: P(III) chlorides are capable to show both electrophilic and nucleophilic properties in the reactions with the same reagent. For example nonsubstituted dialkyl phosphorochloridites react with aldehydes with the formation of polymeric products [27]. In bis(2,2,2-trichloroethyl) - 46a and di(1-chloroethyl) phosphorochloridites 11a the nucleophilicity of P(III) is apparently decreased under the influence of two electron-acceptor 2,2,2-trcihloroethyl or 1-chloroethyl groups. They show only electrophilic properties in reaction with aldehydes forming intermediates 12b and 12a. At the same time P(III) nucleophilicity in ethyl (1-chloroethyl) phosphorochloridite 11c and (1-chloroethyl) ethylphosphonochloridite 11d is probably high enough to react with aldehydes similarly to unsubstituted dialkyl phosphorochloridites. Thus, introducing the chlorine atoms into alkyl groups of dialkyl phosphorochloridites, we managed to direct the reaction towards the formation of primary intermediates 12a-b instead of polymeric products.

EtO(MeCHClO)PCl 11c   Et(MeCHClO)PCl 11d

Similar results were obtained by replacement of ethoxyl group in ethyl phosphorodichloridite by 2,2,2-trcihloroethoxyl one. At interaction of 2,2,2-trichloroethyl phosphorodichloridite 46b with double excess of 2-methylpropanal in the presence of AVE chlorine atoms are gradually substituted by 1-chloroalkoxyl group. The intermediates 11b and 12d are formed. (Table 2.1). After keeping of a reaction mixture for 24 hrs at room temperature in ^{31}P NMR spectrum three resonance signals at δ164, 161 ppm (2:1) and δ179 ppm were observed.

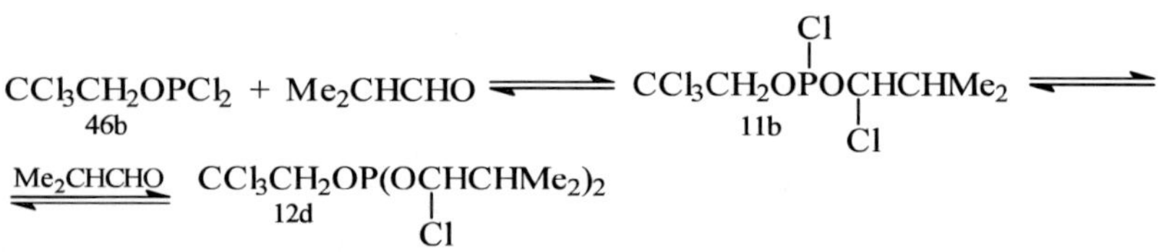

First two of them corresponds to an environment of phosphorus $O_2ClP(III)$ in dialkylphosphoro-chloridites. Therefore they were ascribed to P(III) in intermediate 11b. The third resonance signal corresponds to P(III) in initial phosphorodichloridite 46b. In 6 days intensive resonance signals at δp 142 and 139 ppm (ratio 3:1) were observed in ^{31}P NMR spectrum of a reaction mixture. They correspond to the environment of phosphorus $O_3P(III)$ in trialkyl phosphites. Therefore they were ascribed to P(III) in intermediate 12d. Analogous intermediate 12e was synthesised from acetaldehyde. (Table 2.1)

Intermediates 12k-l containing two non-substituted alkoxyl groups at P(III) were obtained by an indirect method: by interaction of compound 10c with corresponding alcohol in the presence of tertiary amine (Section 5). They were found to be stable only in diluted ether solutions at 0 -5 °C. At removal of solvent or when the temperature of a solution was allowed to reach 20-25 °C exothermal reaction proceeded with the formation of a viscous liquid. In NMR spectra were not observed resonance signals corresponding to the $Me_2CHCH(Cl)OP(III)$ fragment. Intermediates 12k-l wereoxidized into phosphates 65i-g (Section 3).

In chloral carbonyl oxygen possesses lowered nucleophilicity under the influence of three electron-acceptor chlorine atoms. Under the conditions of acid catalysis P(III) chlorides showing electrophilic reactivity do not react with chloral [1-2]. The secondary chlorophosphines and chloral react with the formation of products of betaine structure [62-63]. We proposed that activated by tertiary amine molecule of chloral will react with such strong electrophile as PCl_3. Really, under mild conditions (0 -5 °C) chloral reacts with PCl_3 in the presence of catalytic amounts of tertiary amine (1:1:0.03) with the formation of primary intermediate 10d.

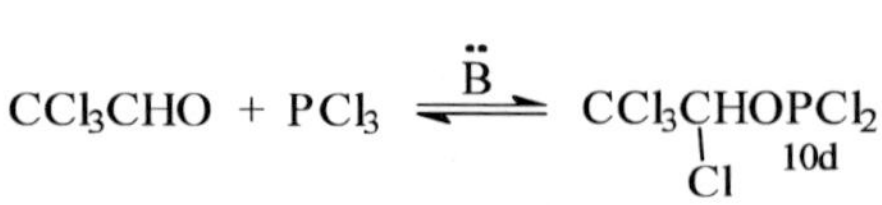

In 5 hrs after mixing of the reagents three resonance signals 200 (PCl_3), 184 and 164 ppm (1:8:1) were observed in ^{31}P NMR spectrum. Second resonance signal corresponds to environment of phosphorus $OCl_2P(III)$ in alkyl phosphorodichloridites. Therefore it was ascribed to P(III) in intermediate 10d (Table 2.1). Other P(III) chlorides showing electrophilic reactivity did not show signs of interaction with chloral. The structure of 10d was confirmed by its NMR spectra and by transformation into stable dichlorophosphate 65l.

The catalysis mechanism of reactions of P(III) chlorides with aldehydes in the presence of tertiary amines or AVE was not specially studied. However, it is possible to assume that as a result of soft interactions of nucleophile with carbonyl carbon nucleophilic properties of carbonyl oxygen atom increase. Carbonyl group is activated and its electrophilic centre is simultaneously covered from the addition of second aldehyde molecule. Activated by nucleophile aldehyde molecule 47 interacts with P (III) chloride with a charge division 48 or via synchronous cyclic transfer of electrons 49. The primary intermediate is formed and the catalyst is regenerated.

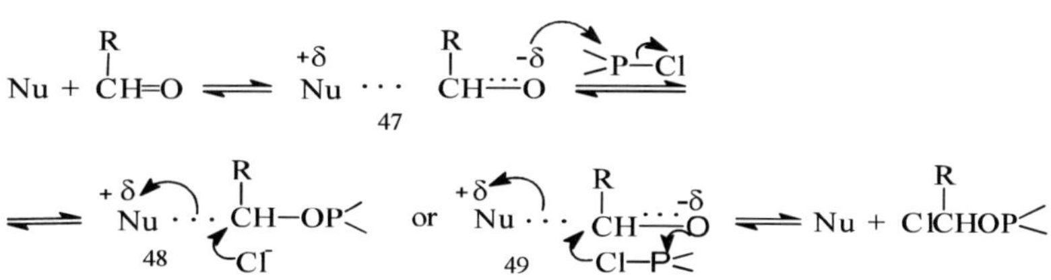

3. Chemical Properties of the Primary Intermediates

Primary intermediates of reactions of P(III) chlorides with aldehydescan be divided into two types: the compounds containing active chlorine atoms at P(III) 10-11, and not containing them, but having fragment $RCH(Cl)OP(III)O_2$12. Compounds 10 are usually stable. They can be distilled *in vacuo* and stored for a long time in the sealed ampoule. The second ones are less stable and decompose into initial P(III) chloride and aldehyde especially at heating. It can be assumed, that such aprotonic nucleophiles as alkene oxides, acetals, trialkyl orthoethers, trialkyl phosphites will interact with the P(III) chloride formed shifting the equilibrium towards decomposition of the intermediate 12. The intermediates of the first type 10 and 11 can interact with them directly.

3.1. Interaction of Intermediates 12with Ethylene Oxide, Acetal, Trialkyl Orthoformates and Trialkyl Phosphites

Oxiranes (alkene oxides) react with all P(III) chlorides with cycle disclosure [1-2, 8, 45-47]. The coordination number of phosphorus atom does not change. Depending on the reagents ratio it is possible to substitute gradually all the chlorine atoms at P(III) by 2-chloroethoxyl group.

$$R_nPCl_{3-n} + (3-n)\,\underset{\diagdown O \diagup}{CH_2\text{—}CH_2} \longrightarrow R_nPOCH_2CH_2Cl)_{3-n}$$

Acetals, ketals, orthoethers of carboxylic acids react with P(III) chlorides with C-O bond cleavage. The reactions are usually result in the formation of P(IV) derivatives. P(III) chlorides show electrophilic properties and a primary process is the replacement of chlorine atoms at P(III) by alkoxyl groups. Depending on a ratio of initial reagents it is possible to substitute all chlorine atoms at P(III). That has been proved by isolation of substitution products and their identification by a chemical methods or by means of ^{31}P NMR [33].

$$R^1_nPCl_{3-n} + (3-n)\,R^3CH(OR^2)_2 \longrightarrow R^1_nP(OR^2)_{3-n} + (3-n)\,R^3CH(OR^2)Cl$$

$$R_nPCl_{3-n} + (3-n)\,HC(OR^2)_3 \longrightarrow R^1_nP(OR^2)_{3-n} + (3-n)\,ClCH(OR^2)_2$$

Compounds 12f-i appeared to be the most suitable objects for studying among the intermediates 12. The results obtained at their interaction with above-mentioned nucleophiles can be presented by the generalizing scheme [57]:

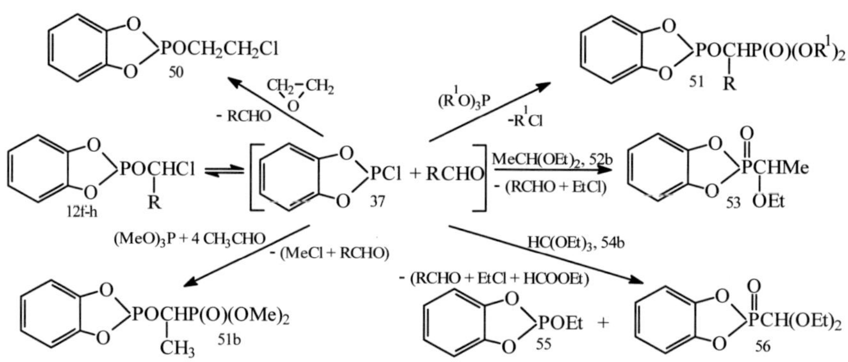

At mixing the compound 12f with ethylene oxide in the ratio 1:1 at +5°C very weak thermal effect (rise in temperature by 2 °C) is observed. In ^{31}P NMR spectrum of reaction mixture a new resonance signal at δp129 ppm is found out and its intensity is slowly growing. Only after 145 hours this signal became dominant.

Distillation *in vacuo* of crude reaction product afforded pure catechol 2-chloroethyl phosphite 50. The structure of 50 was confirmed by ^{1}H and ^{31}P NMR spectra (Table 3.1) and by its synthesis from CCP. The reaction between CCP and ethylene oxide proceeds with heat evolution.

In 24 hrs in ^{31}P NMR spectrum only one resonance signal at δp129 ppm corresponding to P(III) was observed. Constants and ^{1}H NMR spectra data of 50 obtained by different methods appeared to be identical.

Table 3.1. ^{1}H, ^{13}C and ^{31}P NMR data for phosphoryl compounds $R^1 R^2 P(O)R^3$

№	R^1	R^2	R^3	^{m}E	δ, ppm (J_{HH} and J_{PH} , HZ)						
					E – 1; 2	E - 3	E – 4;5	E - 6	E - 7	E - 8	E- 9;10
1	2	3	4	5	6	7	8	9	10	11	12
1	1 MeO	3 6 CH—Me | 4 OMe	7 8 9 $OCHCHMe_2$ | Cl	^{1}H	3.70 d, 3.65 d (10.0);	4.60 m	3.40 s	1.41 dd (7.0,18.0)	6.20 m	1.92 m	1.00 d (6.3);
2	MeO	3 6 CH—Me | 4 OMe	7 8 9 10 $OCHCH_2CH_2Me$ | Cl	^{1}H	3.70 d, 3.65 d (10.0)	4.70 m	3.37 s;	1.22 m	6.11 m	1.92 m	1.42 m; 0.83 t (6.3)
				^{13}C	56.9 d (8.0)	71.7 d (168.0), 72.0 d (162.0), 72.3 d (166.0)	50.8 s; –	13.1 d (12.0)	86.9 d (10.0)	40.6 s	15.8 s; 11.2 s
3	MeO	3 6 CH—Me | 4 OMe	OMe	^{1}H	3.50 d (10.0)	4.50 m	3.20 s; –	1.50 dd (7.0,18.0)	–	–	–
4	MeO	3 4 $CH(OMe)_2$	7 8 OCHMe | Cl	^{1}H	3.56 d, 3.57 d (10.7,10.5)	4.42 d, 4.40 d, 4.35 d, 4.37 d (7.3, 7.4, 6.6,	3.23 m	–	6.05 m	1.54 d 1.56 d (5.1, 5.6)	–
				^{13}C			56.31 d,	–	84.74 d,	29.92 d	
					52.92 d, 52.88 d (6.7, 6.7)	6.6) 100.69d, 99.90 d (206.9, 211.4)	56.16 d, 55.73 d 55.68 d (11.7,11.2, 11.5, 11.5)		84.41 d (6.5, 8.0)	27.24 d (7.0, 6.0)	

Table 3.1. (Continued)

№	R^1	R^2	R^3	^mE	δ, ppm (J_{HH} and J_{PH} , HZ)						
					E – 1; 2	E - 3	E – 4;5	E - 6	E - 7	E - 8	E- 9;10
1	2	3	4	5	6	7	8	9	10	11	12
5	MeO	$CH(OMe)_2$	7 8 9 $CHCHMe_2$ | Cl	^1H	3.89, 3.88 d (11.0, 11.0)	4.67 d 4.73 d (6.2, 7.3)	3.53 m	–	6.09 m	2.20 m	1.07 m –
				^{13}C	53.04 d 52.95 d (7.2, 7.2)	100.9 d 99.9 d (206.6, 211.9)	50.50 – 56.25 m	–	92.65 d 92.19 d (7.5, 9.3)	36.51d 36.32 d (5.1, 6.7)	16.61 s 16.29 s –
6	1 $MeCH_2$ O	3 4 5 $CH(OCH_2M$ $e)_2$	7 $OCHCCl_3$ | Cl	^1H	3.85 quintet (7.5,7.5) 0.85 m	4.35 d (7.5)	3.30 k (7.5); 0.85 m	–	6.0 d, 6.2 d (8.3)	–	–
7	1 MeO	3 4 $CH(OMe)_2$	1 OMe	^1H	3.61d (11.0)	4.67 d (5.0)	3.26 s; –	–	–	–	–
8	2 1 $MeCH_2$ O	3 4 5 $CH(OCH_2M$ $e)_2$	OCH_2Me	^1H	4.1 quintet (7.0,7.0); 1.17 t (7.0)	4.6 d (5.0)	3.70 k, 3.63 k (7.0); 1.10 t (7.0)	–	–	–	–
9	2 1 CH_3CH_2 O	3 4 5 $CH(OCH_2M$ $e)_2$	7 8 9 10 $OCHCH_2CH_2CH$ $_3$ | Cl	^1H	4.12 quintet (7.5,7.5); 1.31 t (7.5)	4.72 d, 4.53 d (5.0)	3.72 m; 1.17 t (7.5)	–	6.12 m	1.98 m	1.48 m; 0.91 t (7.5)
10	8 7 CH_3CH O | Cl	3 4 CHOCHCl |6 |5 Me Me	CH_3CHO | Cl	^1H	–	3.95 m	5.7 k (5.0); 1.6 m	1.20 dd, 1.15 dd, (7.0,18.0)	6.15 m	1.6 m	–

№	R^1	R^2	R^3	^mE	δ, ppm (J_{HH} and J_{PH} , HZ)						
					E – 1; 2	E - 3	E – 4;5	E - 6	E - 7	E - 8	E-9;10
1	2	3	4	5	6	7	8	9	10	11	12
11	1 CCl_3CH_2O	3 6 7 $CHCHMe_2$ | 4 5 8 $OCH{=}CMe_2$	CCl_3CH_2O	^1H	4.57 m; –	3.77 dd (7.5, 7.5)	5.99 s; –	2.2 dheptet (7.5, 7.5)	1.1 d (7.5)	1.5 s, 1.6 s	–
12	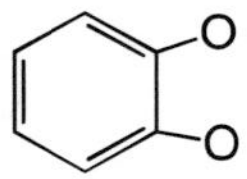		7 8 OCH_2CH_2Cl	^1H	7.05 m	–	–	–	3.8 dt (6.5, 6.5)	3.5 t (6.5, 6.5)	–
13	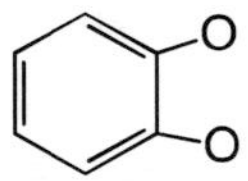		3 6 CHMe | 4 5 OCH_2Me	^1H	6.78 m	4.0 dk (3.5, 7.0)	3.47 k (7.0); 0.88 t (7.0)	1.43 dd (7.0,18.5)	–	–	–
14	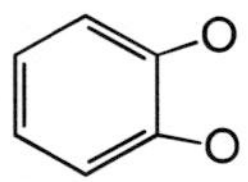		7 8 OCH_2CH_3	^1H	6.70 m	–	–	–	3.6 quintet (7.0,7.0)	t (7.0)	–
15	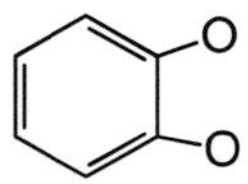		3 4 5 $CH(OCH_2Me)_2$	^1H	6.7 m	4.7 d (5.0)	3.70 k (7.0); 1.2 t (7.0)	–	–	–	–
16	1 MeO		OMe	^1H	3.51 d (10.0)	3.70 m	6.70 m	1.4 dd (7.0,16.0)	–	–	–
		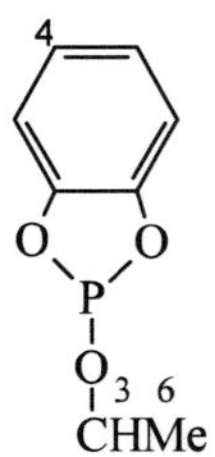 									

Table 3.1. (Continued)

№	R^1	R^2	R^3	^mE	δ, ppm (J_{HH} and J_{PH}, HZ)						
					E – 1; 2	E - 3	E – 4;5	E - 6	E - 7	E - 8	E- 9;10
1	2	3	4	5	6	7	8	9	10	11	12
17	1 MeO	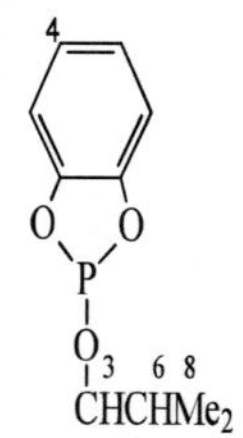	OMe	^1H	3.50 d 3.55 d (10.0)	3.75 m	6.70 m	1.75 m	–	0.70 d, 0.63 d (6.3)	–

In the sixtieth years of last century Birum has published more than ten patents, where reactions of activated by trialkylphosphites carbonyl compounds with P(III) chlorides were described [1, 53]. We proposed that, if to mix intermediates 12 with trialkyl phosphate, the latter will react with the intermediates decomposition products as it occurs in three-component system described by Birum. At addition of trimethyl phosphite to intermediate 12f a small exothermic effect is observed. In 24 hrs only two doublets of approximately identical intensity (δp 23 and 140 ppm) were observed in ^{31}P NMR spectrum corresponding to an environment of phosphorus atoms $O_3P(IV)C$ and $O_3P(III)$ in the product 51a. (Table 3.1).

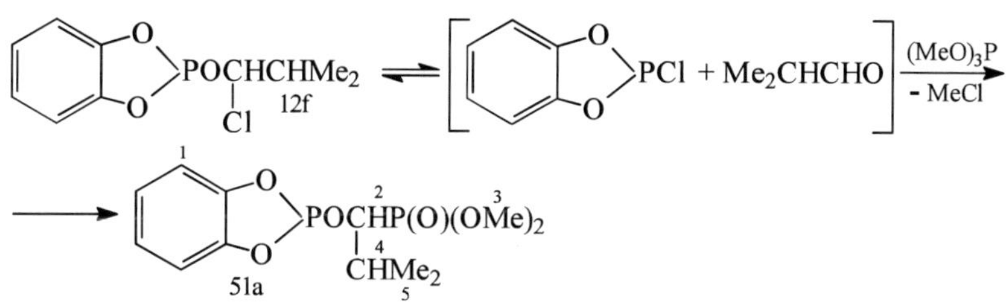

It is necessary to notice that compounds 51 can be also formed directly from intermediates 12f-i as a result of bimolecular nucleophilic substitution.

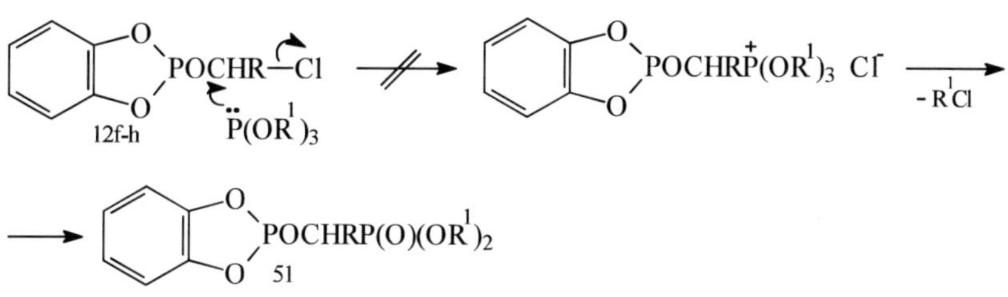

To exclude such a possibility we performed the reaction of intermediate 12f with trimethyl phosphite in the presence of four-fold excess of acetaldehyde. Compound 51b was obtained in a good yield. It does not contain aldehyde fragment of intermediate 12f andthat testifies to its initial decomposition into catecholchlorophosphite and 2-methylpropanal and further interaction of acetaldehyde activated by trimethyl phosphite with P(III) chloride 37.

Interaction of intermediates 12f-i with acetals was studied by the example of diethyl acetal of acetaldehyde. At mixing any of compounds 12f-i with small excess of acetal the insignificant thermal effect was observed. In 48 hrs removal the volatile components and distillation *in vacuo* of crude product afforded pure compound 53. Its constants are close to those described in literature [64]. In ^{1}H NMR spectrum of volatile products condensed in a trap with liquid nitrogen resonance signals corresponding to the aldehyde protons were observed.

Interaction of intermediates 12f-i with triethyl orthoformate 54b also proceeded with insignificant thermal effect. In 96 hrs in ^{31}P NMR spectrum of reaction mixture intensive resonance signal at δp 128 ppm and weak onces at δp 135 and 46 ppm were observed. They correspond to phosphorus atoms in catechol ethyl phosphite 55 and catechol diethoxymethylphosphonate 56, respectively. By distillation *in vacuo* pure compounds 55 and 56 were obtained. Their constants are close to those described in literature [65].

It should be noted that interaction of individual catechol chlorophosphite with ethylene oxide, acetal, trialkyl orthoformate, with aldehydes activated by trialkyl phosphites, proceeds with considerable thermal effect. And the above described reactions of intermediates 12f-i with these nucleophils proceeded with weak exothermic effect. We believe that decomposition of compounds 12f-i into catechol chlorophosphite and aldehyde is a rate-limiting stage of these reactions.

3.2. Reactions of Intermediates10b-C with Acetals and Trialkyl Orthoformates

Intermediates 10b-c in contrast to compounds 12f-i have active chlorine atoms at P(III). Therefore we expected that their reactions with acetals and orthoethers will proceed with significant exothermic effect. On the other hand at stages of substitution of chlorine atoms at P(III) by alkoxyl group unstable intermediates 12k-l should be formed. They can partially decompose into dialkyl phosphorochloridite and aldehyde and then can form oligomer poroducts [27]. To reduce the contribution of decomposition of compounds 12k-l (a route a), the reaction was performed at cooling and intermediate 10b-c was added to an excess of acetal or orthoether. At cooling and dropwise addition of intermediate 10b-c to the two-fold excess of acetal 52a intensive reaction was observed. Two fractions and viscous residue were obtained by distillation *in vacuo.* The distillation residue was not specially studied. At interaction of ordinary dialkyl phosphorochloridites, for example diethyl phosphorochloridite with acetal the non-distilled residue is not formed practically. Therefore we assume that the distillation residue is a result of further reactions of decomposition products of intermediate 12f (route a). Recurring distillation of fractions allowed to isolate pure products 58 and 59 [43].

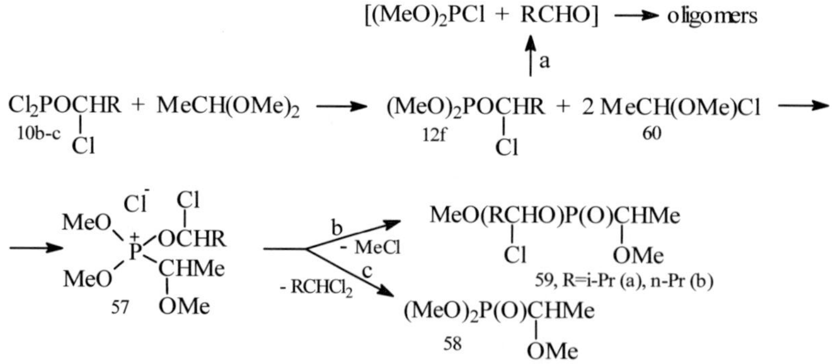

The same low-boiling compound – dimethyl 1-methoxyethylphopshonate 58 was obtained from both intermediates 10b-c. The structure of methyl (1-chloroalkyl)-1-methoxyethylphosphonate 59 was ascribed to the high-boiling product. (Table 3.1)

The formation of two products 58 and 59 is obviously can be explained by stabilisation of intermediate quasiphosphonic salt 57 by two routes: with elimination methyl chloride (route b) and alkylidene dichloride (route c).

We also found out that at dropwise addition of intermediates 10b-c to 1,5-2,0-fold excess of trialkyl orthoformate 54 the exothermic reaction proceeds. Distillation of reaction mixture provided two fractions. Distillation residue was much less than in the case of reaction of compounds 10b-c with acetals. Probably, it is due to more electrophilic reactivity of chloroformals 61 in comparison that of α-chloroethers 60. The contributions of routes (b-c) are increased in comparison with that of route (a). In ^{31}P NMR spectrum of the first fraction three resonance signals at δp15.1, 14.9 (a ratio 2:1) and 16.4 ppm are observed. The ratio of total intensity of the first two signals to intensity of the third signal makes up 1:1. Slow distillation using Vigreux column of 15 cm length made it possible to isolate pure compound 63 [43]. Its ^{31}PNMR spectrum contains resonance signal at δ 16.4 ppm. (Table 3.1).

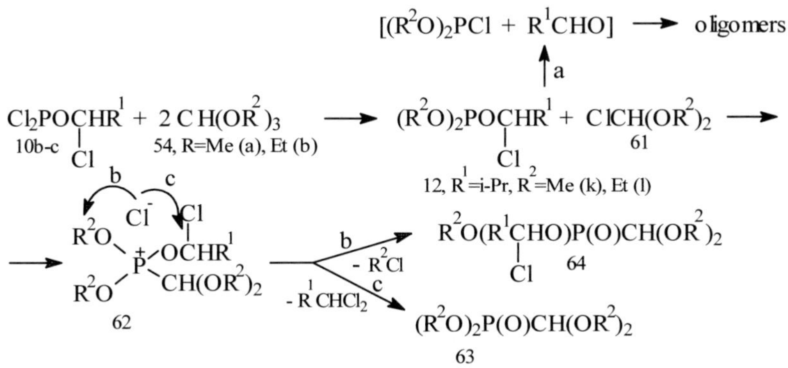

The product 64 was obtaned by distillation of the second fraction using the same Vigreux column. In ^{31}P NMR spectrum two resonance signals at δ_P 15,1 and 14,9 ppm (a ratio 2:1) are observed. ^{1}H, ^{13}C NMR spectra data of compounds 63 and 64 are presented in Table 3.1.

The formation of two products 63 and 64 in this reaction is apparently possible to be explained by stabilization of intermediate quasiphosphonic salt 62 in two routes: with elimination alkyl chloride (route b) and alkylidene dichloride (route c).

There is another way of formation of the compounds 58 and 63: intermediates 10b-cdecompose into phosphorus trichloride and aldehyde, and interaction of PCl_3 with acetals or orthoethers of carboxylic acids results in their formation.

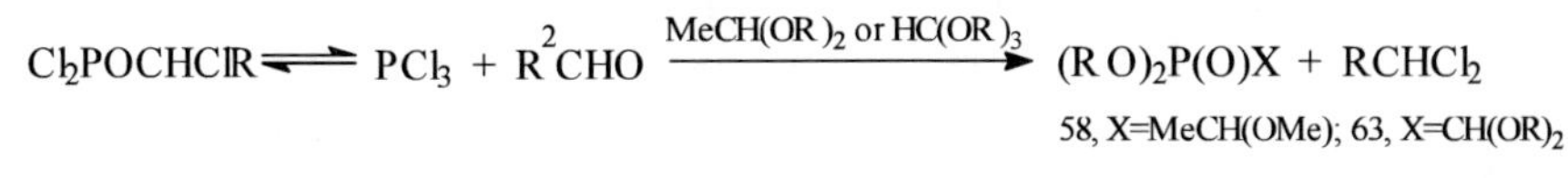

However, even if such way is realised, its contribution is apparently insignificant and these compounds are mainly formed as a result of direct interaction of intermediates 10b-c with acetals and orthoethers. Firstly, compounds 10b-c are incomparably more stable than intermediates 12: they can be stored in a pure state in the sealed ampoule. Secondly, they have active atoms of chlorine at P(III) and as well as other P(III) chlorides react with acetals and otrhoethers with evolution of a considerable amount of heat, while intermediates 12 interact with these nucleophiles rather sluggishly. Thirdly, alkylidene dichloride was found among the volatile products condensed in a trap with liquid nitrogen. And principle possibility of its formation as a result of 1-chloroalkyl group elimination at stabilization of intermediate quasiphosphonic salt was demonstrated by the reaction of intermediate 12f with α-chloroether.

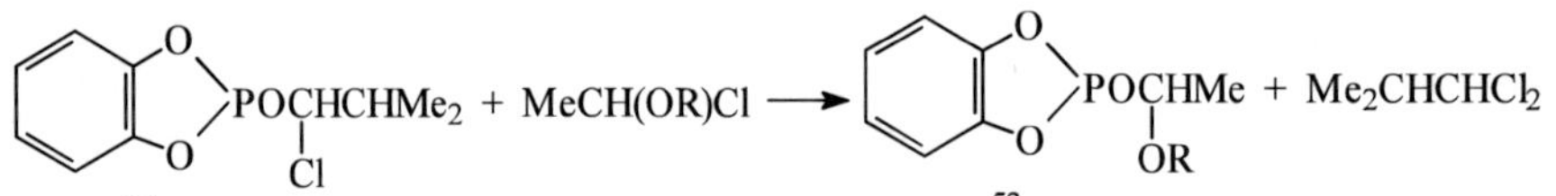

12j 53

The compound 53 synthesized by this reaction and obtained by interaction of intermediates 12f-h with acetals has very close NMR spectral data (Table 3.1). Fourthly, principle possibility of formation of compounds 12k-l which are intermediates in reactions of orthoethers and acetals with 10b-c, was demonstrated by their synthesis using the reactions of the intermediates 10b-c with alcohols in the presence of the base. By this method intermediate 12b was synthesized as well [66].

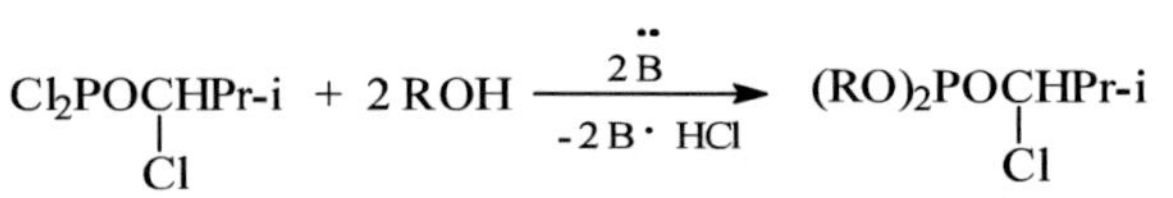

12f, R=Me; 12g, R=Et; 12b, R=CCl_3CH_2

It has appeared that only the substance 12b is stable at a room temperature and it was also synthesized by direct interaction of bis(2,2,2-trichloroethyl)phosphorochloridite with aldehyde.

^{31}P and ^{1}H NMR spectra data of compound 12b synthesized by different methods are identical (Table 2.1).

Compounds 12k-lare stable only in highly-diluted ether solution at 0-5°C. In ^{31}P NMR spectra of these compounds only one resonance signal at δp 140 ppm is observed corresponding to an environment of phosphorus $O_3P(III)$ in trialkyl phosphites. Their structure was also proved by their oxidation to the corresponding phosphates (Section 3.3).

Thus, one can conclude that there is a principle difference in interaction of intermediates synthesized from catecholchlorophosphite 37 and of PCl_3 with acetals and trialkyl orthoformates. The first ones slowly decompose into aldehyde and chloride 37 and then the latter reacts with these nucleophiles. The second ones contain active chlorine atom at P(III). They similarly to all P(III) chlorides interact directly with acetals and orthoethers.

3.3. Oxidation of Primary Intermediates

The oxidation reaction was used as one of the basic chemical methods of proving the structure of primary intermediates. As it follows from the content of a section 2 the phosphorus atoms in the compounds 10-12 synthesized have the substituents with different donor-acceptor properties. So, it was necessary to choose such oxidizers which would oxidize both electrophilic and nucleophilic P(III) derivatives.

The analysis of literature data on oxidation of various P(III) derivatives allowed us to assume that dimethylsulfoxide and tert-buthylhypochlorite can be suitable for oxidation of primary intermediates. They are available and easily and safely to be used. Oxidation can be performed under conditions at which compounds 10-12 are stable and they can olso oxidize both electrophilic and nucleophilic P(III) derivatives.

It was experimentally confirmed that both oxidizers are applicable for oxidation of primary intermediates.

Oxidation was performed under mild conditions: in solvent and in the cold. In all cases stable products of oxidation 65 were isolated. A mixture of intermediate 10a with acetaldehyde trimer was also subjected to oxidation since they are not separated by distillation (Section 2).

The oxidation product 65a has considerably higher boiling temperature than paraldehyde and it was isolated by distillation *in vacuo.* Intermediates 12k-l, obtained by the action of alcohols on compound 10c in the presence of the tertiary amine were oxidized in strongly diluted ether solution just after removal of amine salt. Constants and data of the elemental analysis for substances 65 are presented in Section 5. The structure of the compounds 65 was confirmed by ^{1}H and ^{31}P NMR spectra (Table 3.2).

Oxidation of intermediates can be described by the following scheme [43]:

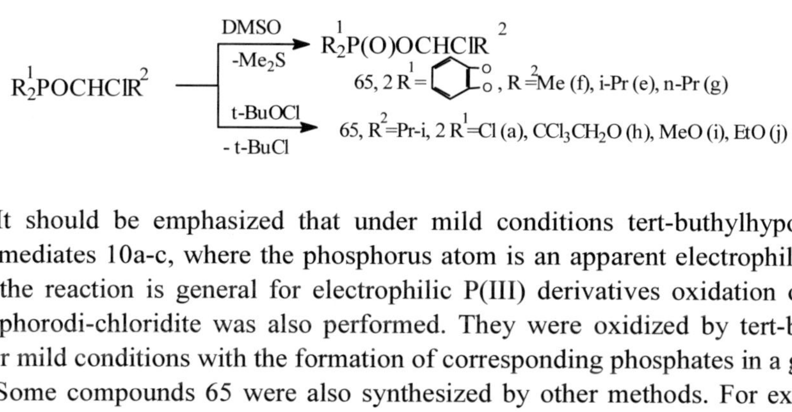

It should be emphasized that under mild conditions tert-buthylhypochloride oxidizes intermediates 10a-c, where the phosphorus atom is an apparent electrophile. To demonstrate that the reaction is general for electrophilic P(III) derivatives oxidation of PCl_3 and ethyl phosphorodi-chloridite was also performed. They were oxidized by tert-buthylhypochlorite under mild conditions with the formation of corresponding phosphates in a good yield.

Some compounds 65 were also synthesized by other methods. For example, substances 65k-mwere obtained by interaction the phosphates 65a and 65d with methanol in the presence of tertiary amine.

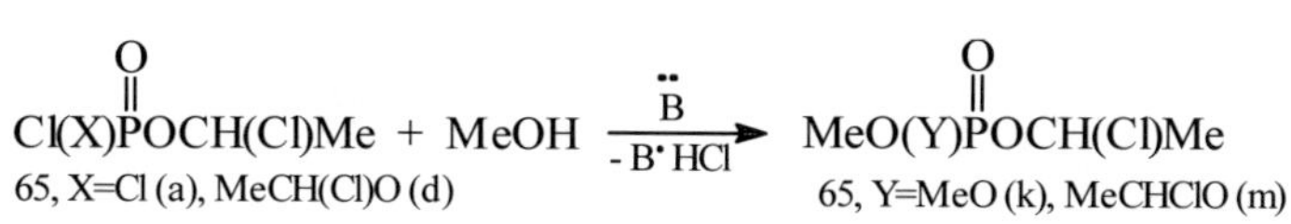

Compound 65e along with corresponding chlorophosphate 66 is formed at interaction of intermediate 12f with PCl_5 (Section 3.4).

3.4. Interaction of Intermediates 12f-Iwith Phosphorus Pentachloide and Chlorine

Trialkyl phosphites 19 react with halogens, in particular with chlorine, with the formation of dialkyl chlorophosphates [37, 67].

$$(RO)_3P + Cl_2 \longrightarrow [(RO)_3PCl_2 \rightleftharpoons (RO)_3\overset{+}{P}Cl\ Cl^-] \longrightarrow (RO)_2P(O)Cl + RCl$$

Phosphorus pentachloride can serve as source of chlorine. At interaction with trialkylphosphites 19, PCl_5 is reduced into PCl_3 and the dialkyl chlorophosphates are the second basic reaction products [67-68].

$$(RO)_3P + PCl_5 \longrightarrow (RO)_2P(O)Cl + PCl_3 + RCl$$

When an excess of trialkyl phosphite is used the process is complicated by its disproportionation under the action of PCl_3 [69].

$$3\,P(OR)_3 + PCl_5 \xrightarrow[-RCl]{} PCl_3 + (RO)_2P(O)Cl + 2\,P(OR)_3 \longrightarrow (RO)_2P(O)Cl + 3\,(RO)_2PCl$$

Other processes complicating this reaction are not described in the literature.

Intermediates 12f-iare full phosphites. Therefore we proposed that they will interact with PCl_5 and Cl_2 similarly to trialkyl phosphites. The chlorophosphate 66 should be the main product of reaction (route a).

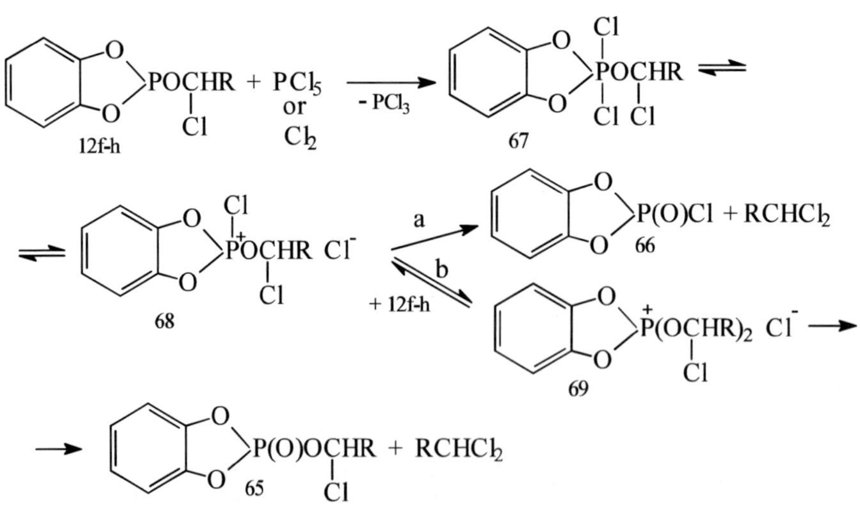

Table 3.2. ^{1}H and ^{31}P NMR data for phosphates R^1 $R^2P(O)OCHClR^3$

N comp	R^1	R^2	R^3	δ, ppm (J_{HH} and J_{PH} , HZ)						$δ_P$ ppm
				H-1	H-2	H-3	H-4	H-5	H-6	
1	2	3	4	5	6	7	8	9	10	11
65a	Cl	Cl	Me	6.30 sectet (10.0, 5.0)	1.80 d (5.0)	-	-	-	-	5.0
65b	Cl	Cl	n-Pr	6,25 dt (5.0, 6.3)	1.88 dt (5.0, 6.3)	1.38 sectet (6.3; 6.3)	0.9 t (6.3)	-	-	6.0
65c	Cl	Cl	$CHMe_2$	6.30 dt (10.0, 5.0)	1.9 m	-	0.95 d	-	-	6.0
65d	Cl	$OCHClCH_3$	Me	6.28 m	1.79 t	-	-	-	-	2.42, 1.20 0.70 (1:0. 5:0.5)
65e	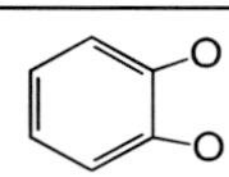		$CHMe_2$	6.12 dd (8.0,5.0)	2.05 m	-	0.98 d (6.0)	7.0 m	-	10.6
65f	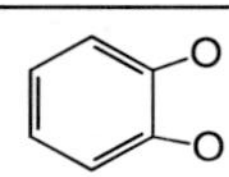		Me	6.12 dk (9.0,5.0), 1.5 d (5.0)	-	-	-	-	-	11.0
65g	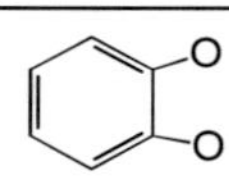		n-Pr	6.23 dt (8.0,5.5)	1.90 m	1.5 m	0.93 t (6.5) 7.0 m	7.0 m	-	11.0
65h	CCl_3 CH_2O	5 CCl_3 CH_2O	2 4 $CHMe_2$	5.92 dd (9.0,5.0)	2.12 m	-	0.99 d (6.3)	4.55 d (5.0)	-	-6
65i	MeO	MeO	$CHMe_2$	5.83 dd, (8.75,5.0)	2.05 m	-	0.99 d (6.3)	3.70 d, 3.74 d (12.5)	-	0

Table 3.2. (Continued)

N comp	R^1	R^2	R^3	δ, ppm (J_{HH} and J_{PH}, HZ)						δ_P ppm
				H-1	H-2	H-3	H-4	H-5	H-6	
1	2	3	4	5	6	7	8	9	10	11
65j	$MeCH_2O$	6 5 $MeCH_2O$	$CHMe_2$	5.85 dk, (9.0,6.3)	1.42 d (6.3)	-	-	4.00, 4.04 quintet	1.2t (7.5)	-3
65l	Cl	Cl	CCl_3	6.52 d (11.3)	-	-	-	-	-	8
65k	MeO	MeO	Me	5.85 dk, (9.0,6.3)	1.42 d (6.3)	-	-	3.40 d, 3.44 d (12.0)	-	-1
65m	MeO	2 1 CH_3CHClO	Me	6.05 m	1.62 m	-	-	3.61 d, (12.0) 3. 63 d, (15.0)	-	-3.89, -4.86, -5.14, (2:1: 0.8)

Chlorophosphate 66 appeared not to be only one phosphorus-containing product of reaction. Along with it full phosphates 67, i.e. oxidation products of intermediates 12f-i are formed (a route b). The ratio of compounds 66 and 65 depends on the order of mixing of the reagents. In this sense the use of $PC1_5$ is more suitable. At dropwise addition of intermediate 12f-i tobenzene solution of PCl_5 chlorophosphate 66 is formed only. If to mix the reagents in reverse order, both compounds 66 and 65 are formed with the ratio 1:1. In ^{31}P NMR spectrum of reaction mixture after removal all volatile products *in vacuo* two singlet resonance signals of equal intensity at δp 19,0 and 10.5 ppm are observed. The first fraction of the first distillation consists of chlorophosphate 66

(δp I9 ppm) . The compound 65e was isolated individually by slow distillation of the third fraction (ratio of 66 and 65 made up 1:4) using Vigreux column of 15 cm length. Constants and spectral data of 65e and of a product obtained by oxidation of intermediate 12f are identical (Section 5).

At chlorination by gaseous chlorine it is difficult to perform reverse mixing the reagents. Therefore in all cases chlorine was passed through a methylene chloride solution of intermediate 12f. Chlorophosphate 66 was mainly formed. The ratio of the signals at δp 19.0 and 10.5 ppm made up 10:1. Compound 66 was isolated with the yield of 78 % by distillation using Vigreux column. The main fraction with ratio of these products 1:3 was obtained at the chlorination of intermediate 12g in toluene. Repeated distillation of this fraction with Vigreux column allowed to isolate the fraction containing full phosphate 65f as the main product (a ratio 1:10). In NMR spectra of this fraction the resonance signals characteristic for compound 65f were observed (Section 5). Thus, experimental data on interaction of intermediates 12f-i with phosphorus pentachloride and chlorine show that these reactions are not simple since oxidation also takes place in addition to chlorination.

We propose that at the beginning intermediates 12f-i reduce PCl_5 into PCl_3 and chlorinate themselves transforming through phosphorane 67 into quasiphosphonic compound 68. The latter enters into the ligand exchange with initial full phosphite 12f-i. New quasiphosphonic compound 69 desalkylates into the oxidation product 65e-f. It should be noted that ligand exchange at quasiphosphonic centre is widely known in the organic chemistry of phosphorus [37, 70-71].

Phosphorus pentachloride was also used for the chemical proof of the structure of tri(1-chloroethyl) phosphite 12a. Di(1-chloroethyl) chlorophosphate 65d formed was also obtained via oxidation of phosphorochloridite 11a by DMSO.

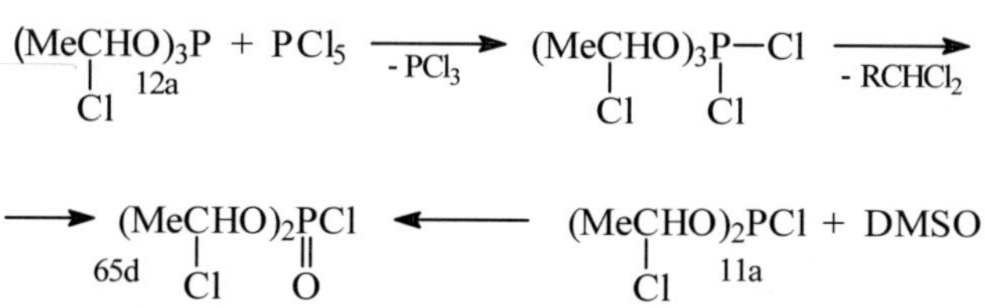

NMR spectral data of compound 65d obtained by different methods, are identical(Section 5).

4. Reactions of the Primary Intermediates with Aldehydes and Detection of the Secondary Intermediates

In section 2 we described the synthesis of compounds 10, 11 and 12, which were called as the primary intermediates in reactions of P(III) chlorides with aldehydes. To confirm, that they are really intermediates in the way of formation of P(IV) derivatives with the fragment >P(O)CHROCH(R)C1, we decided to make them interact with aldehyde. Since the final P(IV) compounds contain a group consisting of two aldehyde fragments , one more intermediate containing the fragment P(III)OCH(R)OCH(R)C1 28 should form between primary intermediates 10,11 and 12 and P(IV) derivatives (Section 1). Therefore, we carried out the research on detecting of compounds, which we call the secondary intermediates.

Intermediate 12a not containing mobile chlorine atom at P(III) was taken as the most suitable one. Addition one molecule of aldehyde to any fragment POCH(Me)C1 of compound 12a can result in secondary intermediate 28d formation. At sufficient nucleophilicity of P(III) the 28a can transform into the P(IV) derivative. We also suspected that there might be a minimum concentration of amine at which its ability to protect the electrophilic centre of the carbonyl group would weakened, although it would still function as a catalyst. This idea was realized in two ways of experiment procedure. The first way assumes the primary formation of intermediate 12a. At 0^{0}C 4.5-fold excess of aldehyde was dropped into the mixture of PCl_3 and tertiary amine in ratio 1:0.03. Intermediate 12a with admixture of 11a was formed (^{31}P NMR data, Section 2). It was allowed the temperature to reach 25 ^{0}C and the reaction mixture to stand for 2 hrs. At this more severe temperature conditions the intermediates 11a and 12a completely transformed into P(IV) derivative. In ^{31}P NMR spectrum a resonance signal at δ_p21 ppm was observed. It corresponds to phosphorus environment O_3CP(IV) in phosphonates. By distillation *in vacuo* the main fraction was isolated which according to elemental analysis and H and ^{31}P NMR spectra has a structure of di(1-chloroethyl)-1-(1-chloroethoxylethylphosphonate 24g (Table 3.1).

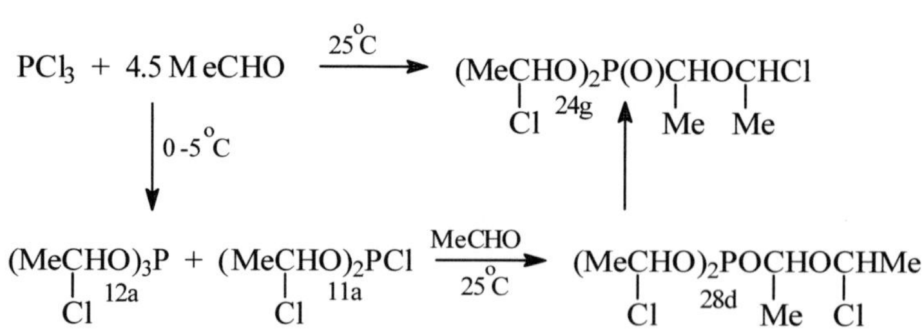

The compound 24g was also synthesized by mixing initial reagents in the same order and ratio, but at the temperature of 25 ^{0}C for 3 hrs. A resonance signal at δ_p21 ppm was observed in ^{31}P NMR spectrum of reaction mixture. The product separated by distilling *in vacuo* turns to be the compound 24g according to ^{1}H and ^{31}P spectra. Thus, the synthesis of P(IV) derivative 24g from PCl_3 and aldehyde was performed through the stage of fixing the formation of intermediates 11a and 12a. It should be noted that intermediates 11a and 12a interact with large aldehyde excess in the same conditions, at which PCl_3 reacts with aldehydes to form directly P(IV) derivatives.

In the experiments described above we failed to fix the formation of secondary intermediate 28d by ^{31}P NMR method. Obviously, P(III) in intermediate proposed possesses enough high nucleophilicity and 28d easily isomerizes into phosphonate 24g. For further decrease of P(III) nucleophilicity in 28 we decided to replace the 1-chloroethoxyl groups at P(III) by the more election-acceptor 2,2,2-trichloroethoxyl ones. Bis(2,2,2-trichloroethyl) phosphorochloridite 46a was used as a P(III) chloride. At the beginning the primary intermediate 12b was obtained from 46a. And then 12b was allowed to react with 2-methylpropanal on ratio 1:1 at 0 ^{0}C.

Under this mild conditions we succeeded to detect the secondary intermediate 28e. Immediately after mixing 2b and 46a in ^{31}P NMR spectrum the new resonance signals at δ_P 132.2 and 24 ppm (in addition to 138.4 and 138.1 ppm in initial 12b) were observed. They were correspond to phosphorus atoms in secondary intermediate 28e and in phosphonate 24h. By allowing the temperature to reach 25 ^{0}C for 0.5 hours the resonance signal at δ_P 132.2 disappeared.

In ^{31}P NMR spectrum a very intensive resonance signal at 24 ppm was observed corresponding to P(IV) in phosphonate 24h in addition to weak resonance signal at δ_P138.4 and 138.1 ppm from phosphorus atom in 12b.

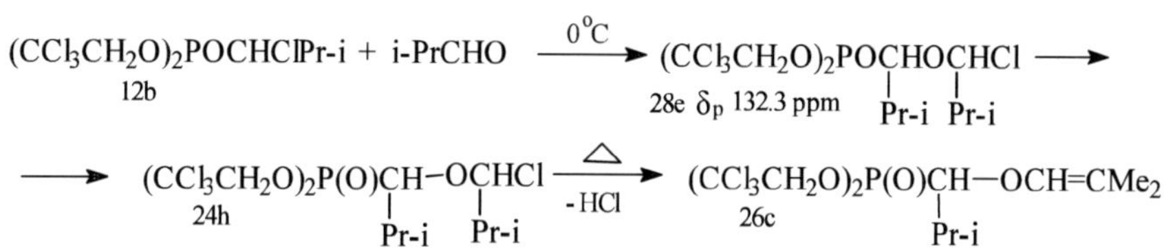

Thus, by ^{31}P NMR method we demonstrated transformation of primary intermediate 12b into secondary one 28e and the latter one into phosphonate 24h.

By carrying out the reaction of phosporochloridite 46a with 2-methylpropanal in the presence of N,N-dietylaniline in ratio 1:3:0.0075 at temperature of 0 ^{0}C it was observed consecutive formation of two intermediates 12b and 28e from chloride 46a and transformation of 28e into phosphoryl compound 24h: the resonance signals at δ_P 163.3, 138.4 and 138.1, 132.2 and 24.0 ppm were observed corresponding to the phosphorus atoms in the compounds 46a, 12b, 28e and 24h respectively.

At increasing the temperature of reaction mixture to 25 ^{0}C within forty minutes the resonance signal at δ_P132.2 disappeared and the intensive resonance signals at δ_P 163.3 and 24 ppm were observed. At distilling the reaction mixture *in vacuo* intensive elimination of HClwas observed. The pure unsaturated phosphonate 26c was isolated.

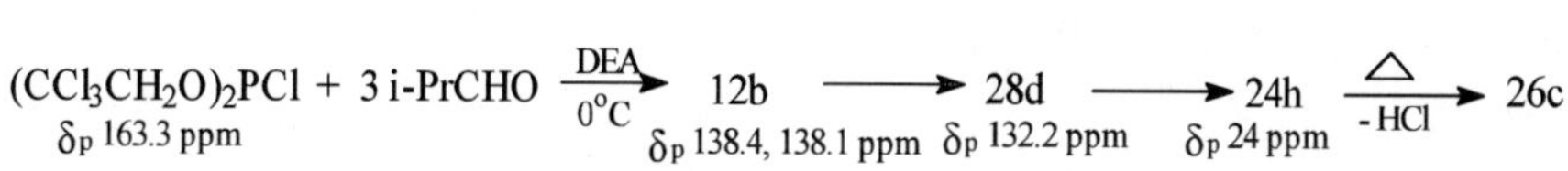

When three-coordinated phosphorus atom in the secondary intermediate 28 is directly bound with two atoms of chlorine: $Cl_2POCHROCHRCl$ 28a, nucleophilicity of P(III)is striongly lowered. It is known that alkylphosphorodichloridites ($ROPCl_2$), being more simple analogues of compounds 28a, are very weak nucleophils. They react with such strong electrophiles as α-chloroethers only in the presence of Lewis acids being the catalysts [72]. In Section 1 it was noted that PCl_3 interacts with aldehydes under mild conditions (70 ^{0}C) with the formation of di-α-chloroethers 16. However their precursors were not detected.

We believe that they are secondary intermediates 28a. We also proposed that there might be a minimum concentration of amine at which its ability to protect the electrophilic centre of carbonyl group would be weakened, although it would still function as a catalyst.

That is why PCl_3 was mixedwith very small amount of tertiary amine and aldehyde was added into this mixture (1: 0.0075:1).

At the very beginning of the experiment low concentration of tertiary amine was created in the reaction mixture. This should facilitate the addintion of the second molecule of aldehyde to the intermediate 10a with the formation of the compound $Cl_2POCHMeOCHMeCl$ 28a.

In 5 hrs after mixing of the reagents the resonance signals at δ_p 219 (PCl_3), 178.1, 177.8, (10a), 169.6, 169.3 ppm were observed in the ^{31}P NMR spectrum. The last two ones were ascribed to the phosphorus atom in the secondary intermediate 28a.

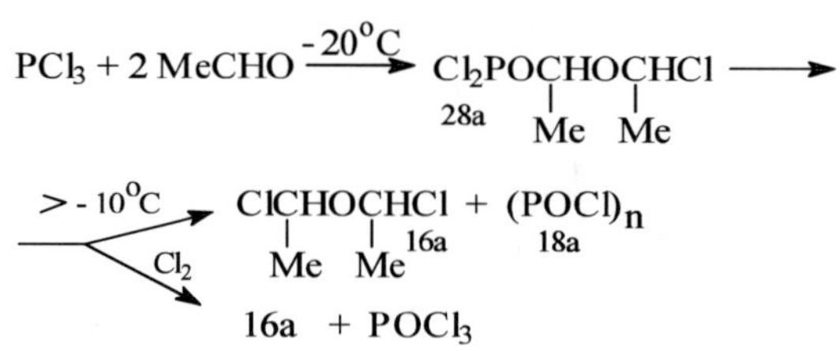

According to integral intensities of the signals the ratio of products 1:10a:28a makes up 22:1:7. The main component of mixture is the initial PCl_3. The secondby content product appears to be the secondary intermediate 28a. The primary intermediate 10a presents only in small amount. In ^1H NMR spectrum there were intensive signals corresponding to protons of compound 28a and weak ones from protons of aldehyde 2a, paraldehyde 45 and initial intermediate 10a (Section 5).

When a ratio of reagents makes up 1:0.0075:2 the overall picture of interaction does not change. In ^{31}P NMR spectrum the same resonance signals arise, but their ratio turns into 10:1:13. The corresponding changes of resonance signal intensities also occur in ^1H and ^{13}C NMR spectra.

The compound 28a turned to be stable only at the temperature below -20 ^{0}C. At the increase of temperature above -10 ^{0}C exothermic reaction occured and one of di-α-chloroethers 16 – 1-chloro -1-(1-chloroethoxy)ethane16a was formed. At the beginning di-α-chloroether 16a was distilled *in vacuo* into a trap with liquid nitrogen and then was separated by distillation.

The phosphorus containing products were not identified. It should be also emphasized that spontaneous transformation of compound 28a into di-α-chloroether 16a justifies its being a direct precursor. Their formation in the reactions of P(III) chlorides with aldehydes was noted by other researches [20-21].

The structure of the intermediate 28a was confirmed chemically. It reacts with chlorine to produce 16a and $POCl_3$. 2-Methyl-1-chloro-1-(2-methyl-1-chloropropoxy)propane16b and 1-chloro-1-(1-chlorobutoxy)butane 16c were synthesized, when 2-metynpronanal and butanal were used as a initial aldehydes.

Thus, in special conditions, namely by creating very low concentration of tertiary amine and keeping low temperature in the reaction medium (not higher -20 ^{0}C) we succeeded in detecting of secondary intermediates 28a in reaction of PCl_3 with aldehyde. They spontaneously transform into 1-chloro-1-(1-chloroalkoxy)alkanes being their direct precursors.

Conclusion

Both hydrogen chloride and tertiary amines or alkyl vinyl ethers(AVE) have been found to catalyse the reaction of P(III) with aldehydes, a reaction studied for over century. A refinement of P(III) chlorides from HCl contaminant was developed using tertiary amines and AVE. Three types of the primary intermediates with different phosphorus environment $OCl_2P(III)$ 10, $O_2ClP(III)$ 11 and $O_3P(III)$ 12 were synthesized. Their structure was confirmed by ^{1}H, ^{13} C and ^{31} P NMR spectroscopy and by transformation into stable species. The electron acceptor substituents at P(III) and electron donor groups in aldehyde fragment stabilize the primary intermediates. Both 11 and 12 revert to P(III) chloride 1 and aldehyde 2 on heating, thus the reactions are reversible. The reactions of 12 with nucleophiles (oxirane, acetals, orthoethers) proceed with a small heat evolution since at the beginning they decompose slowly into the initial compounds. The intermediates 10 and 11 contain chlorine atom at P(III) and they react with nucleophiles directly with large exothermic effect.

By NMR spectroscopy the formation of the secondary intermediates $Cl_2P(O)OCHROCHRCl$ 28a and $(CCl_3CH_2O)_2$ POCHPr-iOCHPr-iCl 28ewas demonstrated, when a very small concentration of tertiary amine was used in reaction. Further their stabilization depends on nucleophilicity of P(III). In 28a P(III) possesses a low nucleophilicity and at higher than – 10 ^{0}C they decompose into 1-chloro-1-(1-chloroalkoxy)alkanes. Therefore we assumed that 28a are their precursors. In secondary intermediates 28 d-eP(III) possesses enough nucleophilicity and they undergo intermolecular Michaelis-Arbuzov isomerization.

5. Description of the Experiments

5.1. Removing of HCl Contaminant from P(III) Chlorides

5.1.1. By Treatment with Tertiary Amines

Commercial phosphorus trichloride was mixed with 15 mole % N,N-diethylaniline and distilled directly into the reaction flask. Pyridine, 2,6-dimethylpyridine, triethylamine and other tertiary amines could be also used.

5.1.2. By Treatment with AVE

Ethyl vinyl ether (4.13 g, 60 mmol) was added dropwise to a stirred catecholchlorophosphite (10 g, 60 mmol) at 10 °C. The α-chloroether and excess of AVE were removed *in vacuo.* Catecholchlorophosphite was distilled directly into the reaction flask. Butyl vinyl and isobutyl vinyl ethers could be also used.

5.2. Reactions of P(III) Chlorides with Aldehydes

5.2.1. Interaction of P(III) Chlorides with Acetaldehyde in the Absence of Catalyst

Acetaldehyde (3.2 g, 73 mmol) was added dropwise to a stirred PCl_3(10 g, 73 mmol) twice treated with N,N-diethylaniline at -5°C. An increase of temperature was not observed. The mixture was allowed to stand for 27 hrs in a refrigerator. ^{31}P NMR (δ, ppm): 220 (PCl_3); ^{1}H NMR (δ, ppm): 1.29 d $^{3}J_{HH}$ 5.1 Hz, 3H, Me; 4.94 k ,$^{3}J_{HH}$ 5.1Hz, 1H, CH; ^{13}C NMR (δ, ppm): 19.89 s, Me, 97.5 s CH (paraldehyde). PCl_3 was removed in vacuo (120 mm) and 2.05 g of paraldehyde was isolated by distillation of crude product.

5.2.2. Reaction of P(III) Chloride with Aldehydes in the Presence of Tertiary Amines

General Procedure

The aldehyde was added dropwise to a stirred mixture of PCl_3 and 1-5% amine at about 5°C. The reaction mixture was allowed to stand for 12-15 hrs in a refrigerator. The volatile products were removed *in vacuo.*

1-Chloro-2-Methylpropyl Phosphorodichloridite 10c

The reaction mixture obtained from PCl_3 (10 g, 73 mmol), 2,6-dimethylpyridine(0,39g, 3,65 mmol)and 2-methylpropanal(5,26g, 73 mmol) was dissolved in dry hexane (15 ml), filtered and concentrated *in vacuo.* Distillation of crude product afforded pure **10c**: 8.17 g (53.3%); b.p. 30 °C (0.01mm), d_4^{20} 1.3029. Anal. Calcd for $C_4H_8Cl_3OP$: C, 22.91; H 3.82; Cl, 50.83; P, 14.79. Found C, 22.74; H, 3.81; Cl, 50.51; P, 14.65.

1-Chlorobutyl Phosphorodichloridite 10b

The reaction mixture from PCl_3 (12.3 g, 90 mmol), butanal (6.46 g, 90 mmol) and N,N-diethylaniline (0.27 g, 1.8 mmol) was distilled in vacuo and 10b was isolated: 10.7 g (56.6%), b.p. 30-32 °C (0.02 mm), d_4^{20} 1.4821. Anal. Calcd for $C_4H_8Cl_3OP$: C, 22.91; H, 3.82; Cl, 50.83; P, 14.79. Found C, 23.13; H, 4.02; Cl, 50.67; P, 14.60.

1-Chloroethyl Phosphorodichloridite 10a

The reaction mixture from PCl_3 (18.36 g, 134 mmol), N,N-diethylaniline (0.4 g, 2.7 mmol) and ethanal (5.9 g, 134 mmol) was distilled after removing of volatile products in vacuo and 12.8 g of mixture of 10b and 45a (3:1 according to ^{1}H NMR) was obtained. NMR data of 10a are presented in Table2.1.

3g of this mixture were heated at 20-90 °C *in vacuo* and volatile products were collected in a trap cooled by liquid nitrogen. ^{1}H and ^{31}P NMR of the trap mixture(δ , ppm) 2.1 d, 9.7 k (2a); 1.29 d, 4.94 k (45a); 220 (PCl_3).

Di(1-chloroethyl) phosphorochloridite 11a

The reaction mixture obtained from PCl_3 (9.33 g, 67.8 mmol), ethanal (7.46 g, 169.5 mmol) and N,N-diethylaniline (0.51 g, 3.39 mmol) gives 13.40 g (87.7%) crude 11a after removing of volatile products *in vacuo*(Table 2.1). The crude 11a was used directly for oxidation by DMSO. By heating 11a*in vacuo* PCl_3, 10a, 2a and 45a were collected in a trap cooled by liquid nitrogen.

Tri(1-chloroethyl) Phosphite 12a

The volatile products were removed in vacuo from the reaction mixture obtained from PCl_3 (12g, 88 mmol), ethanal (17.4 g, 396 mmol) and N,N-diethylaniline (0.66 g, 4.4 mmol). ^1H and ^{31}P NMR spectra of crude 12a were recorded (Table 2.1).

PCl_3, 10a, 2a and 45a were collected in a trap with liquid N_2 by heating 12a*in vacuo.*

On the analogy of12a*1,2,2,2-tetrachloroethyl phosphorochloridite **10d** and 1-chloro-2-methylpropyl diphenyl phosphite **12c*** were prepared from PCl_3 (16.2 g, 118 mmol) chloral (17.38 g, 118 mmol) N,N-diethylaniline (0.35 g, 2.4 mmol) and diphenyl phosphorochloridite 6.45 g (26 mmol), triethylamine (0.053 g, 0.52 mmol), 2-methylpropanal (1.87 g, 26 mmol), respectively. ^1H and ^{31}P NMR data of crude 10d and 12c are presented in Table2.1.

5.2.3. Reaction of P(III) Chlorides with Aldehydes in Presence of EVE

General Procedure

Ethyl vinyl ether and aldehyde were consecutively added dropwise to refined P(III) chloride at 10 °C. The reaction mixture was allowed to stand for 72 hrs in a refrigerator. The volatile products were removed *in vacuo.*

Catechol 1-chloro-2,2-dimethylpropyl Phosphite 12i

The reaction mixture from ethyl vinyl ether (2.5 g, 35 mmol), trimethylacetaldehyde (6.06 g, 70 mmol) and catechol chlorophosphite (6.14 g, 35 mol) was distilled *in vacuo* and 12i was isolated: 6 g (77%); b.p. 74-75 °C (0.02 mm), d_4^{20} 1.2159(Table 2.1). Anal.Calcd for $C_{11}H_{14}ClO_3P$: C, 50.67; H, 5.37; Cl, 13.60. P, 11.88.Found C, 50.52; H, 5.29; Cl, 13.85; P, 11.61.

Catechol 1-chloro-2-methylpropyl Phosphite 12f.

The reaction mixture from EVE (17.7 g, 200 mmol), 2-methylpropanal (50.47 g, 700 mmol) and catechol chlorophosphite (17.79 g, 100 mmol) gives after removing of volatile products *in vacuo* crude**12f** :24.5 g (99.4%)(Table 2.1).

1-Chloro-2-methylpropyl bis(2,2,2-trichloroethyl) Phosphite 12b

From EVE (7.5 g, 104 mmol), 2-methylpropanal (18.7 g, 260 mmol) and bis(2,2,2-trichloroethyl) phosphorochloridite (19 g, 52 mmol) crude **12b** was obtained: 20.1 g (88.9%) (Table 2.1)

Di(1-chloroethyl) 2,2,2-trichloroethyl Phosphite 12e

From EVE (17.2 g, 240 mmol), ethanal 36.7 g, 830 mmol) and 2,2,2-(trichloroethyl) phosphorodichloridite (29.8 g, 119 mmol) crude 12e was obtained 35.63 g (88.5%) (Table 2.1).

5.3. Reactions of Primary Intermediates with Nucleophiles

5.3.1. With Ethylene Oxide

A) Ethylene oxide (5.28 g, 88 mmol) was dropped into catechol 1-chloro-2-methylpropyl phosphite 12f (19.73 g, 80 mmol) at 5 °C. The reaction mixture was allowed to stand 144 hrs at the same temperature. It was distilled *in vacuo* and catechol 2-chloroethyl phosphite was isolated: 13.3 g (76%), b.p. 66-68 (0.03 mm), d_4^{20} 1.3091.

B) Ethylene oxide (5 g, 30 mmol) was very slowly passed through catechol chlorophosphite 37 (5 g, 30 mmol) at 5°C. In 24 hrs by distillation catechol 2-chloroethyl phosphite 50 was isolated: 5.2 g (80%), b.p. 67-69 °C (0.03 mm), d_4^{20} 1.3110 (Table 2.1).

5.3.2. With Trialkylphosphite

A) Catechol 1-chloro-2-methylpropyl phosphite 12f (23.63 g, 96 mmol) was added dropwise to the solution of triethyl phosphite (18.4 g, 110 mmol) in 30 ml benzene. The reaction mixture was allowed to stand 24 hrs at room temperature. By distillation *in vacuocatechol 2-methyl-1-(diethylphosphoryl)propyl phosphite*51c was isolated: 19.6 g (59 %), b.p. 133-134 (0.02 mm), d_4^{20} 1.1933 (Table 3.1). Anal. Calcd for $C_{14}H_{22}O_6P_2$:C, 48.27; H, 6.32; P, 17.79. Found C, 48.01; H, 6.15; P, 17.35.

B) On the analogy of 51c, *catechol 2-methyl-1-(dimethylphosphoryl) propyl phosphite* 51b was obtained from 12f (10.92 g, 44 mmol) and trimethyl phosphite (5.45 g, 44 mmol): 7.8 g (55%), b.p. 135-138 (0.1 mm), d_4^{20} 1.2135. Anal. Calcd for $C_{12}H_{18}O_6P_2$: P, 19.35. Found P, 19.58.

C) *In the presence of acetaldehyde.*12f (10.95 g, 44 mmol) was dropped to the mixture of acetaldehyde (7.78 g, 176 mmol) and trimethyl phosphite (6.6 g, 53 mmol). After removing the volatile products *in vacuo*, distillation of crude product afforded pure *catechol 1-(dimethylphosphoryl)ethyl phosphite*51a: 7.8 g (61%), b.p. 117-118 °C (0.1 mm), d_4^{20} 1.310 (Table 3.1). Anal. Calcd for $C_{10}H_{14}O_6P_2$: C, 41.01; H, 4.79; P, 21.23. Found C, 40.89; H, 4.68; P, 21.15.

5.3.3. With the Acetals

A) 12f (19.173 g, 80 mmol) was added dropwise to 1,1-diethoxyethane (14.18 g, 120 mmol) at room temperature. After 48 hrs standing distillation of reaction mixture afforded pure *catechol 1-ethoxyethylphosphonate*53: 15.49 (71%), b.p. 111 °C (0.02 mm), d_4^{20} 1.2141 (Table 3.1). Anal. Calcd for $C_{10}H_{13}O_4P$: C, 52.63;H, 5.70; P, 13.60. Found C, 52.41; H, 5.58; P, 13.47.

B) 10b (9.06 g, 43 mmol) was dropped to 1,1-dimethoxyethane (7.8 g, 86 mmol) at 20°C. Distillation of reaction mixture afforded *dimethyl 1-methoxyethyl-*

*phosphonate*58: 1.0 g, b.p. 85 °C (9.0 mm) and*1-chloro-2-methylpropyl methyl 1-methoxyethylphosphonate* 59b: 3.56 g (34%), b.p.70-72 °C (0.04 mm), d_4^{20} 14.93 (Table 3.1). Anal. Calcd for $C_8H_{18}ClO_4P$: C, 42.72; H, 8.02; Cl, 14.51; P, 12.7. Found C, 42.80; H, 7.91; Cl, 14.83; P, 13.01.

C) On the analogy of 10b, from 10a (9.6 g, 53 mmol) and 1,1-dimethoxyethane (9.6 g, 106 mmol) *dimethyl 1-methoxyethylphosphonate*58: 0.45 g, b.p. 85 °C (9.0 mm) and *1-chloroethyl methyl 1-methoxyethylphosphonate*59a: 2.16 g (22%), b.p. 51-52 °C (0.04 mm), d_4^{20} 1.1940 (Table3.1) were obtained. Anal. Calcd for $C_6H_{14}ClO_4P$: C, 33.26; H, 6.47; Cl, 16.39; P, 14.31. Found C, 33.10; H, 6.26; Cl, 16.22; P, 14.20.

D) On the analogy of 10b, from 10c (5 g, 24 mmol) and 1,1-dimethoxethane (4.33 g, 48 mmol) *dimethyl 1-methoxyethylphosphonate*58: 0.9 g, b.p. 85 °C (9.0 mm) and *1-chlorobutyl methyl 1-methoxyethylphosphonate*59c: 1.81 g (34 %), b.p. 68-71 °C (0.03 mm) (Table 3.1) were isolated. Anal. Calcd for $C_8H_{18}ClO_4P$: C, 39.26; H, 7.36; Cl 14.51; P, 12.66. Found C, 39.01; H, 7.15; Cl, 14.31; P, 12.35.

5.3.4. With Trialkyl Orthoformate

A) Catechol 1-chloro-2-methylpropyl phosphite 12f (21.8 g, 88 mmol) was dropped into triethyl orthorformate (13.1 g, 88 mmol) at room temperature. The reaction mixture was allowed to stand for 96 hrs. After removing the volatile products *in vacuo*, distillation of crude product afforded *catechol ethyl phosphite*55: 6.7 g. b.p. 92 °C (10 mm), d_4^{20} 1.1925 [97°C(14 mm)], d_4^{20} 1.1954 [89] and *catechol diethoxymethylphosphonate*56a: 5.85 g (28,8 %) b.p. 123-125 °C (0.01) (Table3.1). Anal. Calcd for $C_{11}H_{15}O_5P$: C, 51.16; H, 5.81; P, 12.01. Found C, 51.03; H, 5.76; P, 11.97.

B) 1-chloro-2-methylpropyl phosphosphorochloridite 10c(6.45 g, 31 mmol) was added dropwise to trimethyl orthoformate (7.16 g, 67.5 mmol). The mixture was stirred 0.5 hr at room temperature. Methyl chloride, methyl formate and an excess of trimethyl orthoformate were removed *in vacuo*. Distillacation of crude product afforded*dimethyl dimethoxymethylphosphonate*63a: 0.8 g; b.p. 47-49 °C (0.01 mm) and *1-chloro-2-methylpropyl methyl dimethoxymethyl-phosphonate*64a: 5.3 g (66%) b.p. 93-94 °C (0.01 mm) (Table 3.1). Anal. Calcd for $C_8H_{18}ClO_5P$: C, 36.85; H, 6.91; Cl, 13.62; P, 11.88. Found C, 36.71; H, 6.96; Cl, 13.23; P, 12.01.

C) On analogy with 10c from 10b (11.33 g, 54 mmol) and triethyl orthoformate (16.0 g, 108 mmol) *diethyl diethoxymethyl phosphonate*63b: 1,1 g, b.p. 65-66 °C (0.01 mm), d_4^{20} 1.0531 (127-128 °C (12 mm), d_4^{20} 1.0538 [74] (Table 3.1) and *1-chlorobutyl ethyl diethoxymethylphosphonate*64b: 8.2 g (50%) , b.p. 103-104 °C (0.01 mm), d_4^{20} 1.0978 (Table 3.1) were obtained. Anal. Calcd for $C_{11}H_{24}ClO_5P$: C, 43.64; H, 7.93; Cl, 11.71; P, 10.23. Found C, 43.60; H, 7.83; Cl, 12.10; P, 9.89.

5.3.5. With α-Chlorodiethyl Ether

α-Chlorodiethyl ether (15.1 g, 139 mmol) was dropped into catechol 1-chloro-2-methylpropyl phosphite 12f (34.4 g, 139 mmol). The reaction mixture was allowed to stand for 14hrs at room temperature. It was distilled *in vacuo* and *catechol 1-ethoxy-ethylphosphonate* was isolated 53: 21.57 g (68%), b.p. 113-114 °C (0.01 mm), d_4^{20}1.2150 (Table3.1).

5.4. Oxidation of the Primary Intermediates

5.4.1. With Dimethylsulfoxide (DMSO)

1-Chloroethyl Dichlorophosphate 65a

Benzene solution of DMSO (2.79 g, 36 mmol) was dropped into benzene solution of 1-chloroethylphosphorodichloridite **10a** with admixture of paraldehyde (6.47 g, 36 mmol) at 0-5 ^{0}C. Removing of volatile components and distillation of crude product *in vacuo* gave pure **65a**: 2.13 g (30%), b.p. 30 ^{0}C (0,04 mm), d_4^{20} 1.4849 **(Table 3.2)** Anal. Calcd for $C_2H_4Cl_3O_2P$ C, 12.15; H, 2.02; Cl, 53.92; P, 15.69. Found C. 11.92; H, 1.92; Cl, 53.80; H, 15.63.

1-Chlorobutyl Dichlorophosphate 65b

On analogy with **10a** from DMSO (2.04 g, 26 mmol) and 10b (5.48 g, 26 mmol) 65b was obtained: 2 g (31%), b.p. 40 ^{0}C (0.025 mm) (Table3.2).Anal. Calcd for $C_4H_8Cl_3O_2$ P: C, 21.29; H, 3.54; Cl, 47.18; P, 13.74. Found C, 21.09; H, 3.24; Cl, 47.11; P, 13.72.

1-Chloro-2-methylpropyl Dichlorophosphate65c

On analogy with **10a** from DMSO (16.09 g, 206 mmol) and 10c(51g, 206 mmol) 65c was obtained: 28 g (52%), b.p. 43 ^{0}C (0.01 mm) (Table3.2) Anal. Calcd for $C_4H_8Cl_3O_2P$: C, 21.29; H, 3.54; Cl, 47.18; P, 13.74. Found C, 21.01; H, 3.50; Cl, 47.15; P, 13.68.

Catechol 1-chloroethyl Phosphate 65f

On analogy with 10a from DMSO (11.8 g, 151 mmol) and 12g (33.0 g, 151 mmol) 65f was isolated: 13.41 (38%), b.p. 95-97 ^{0} C (0.03 mm), d_4^{20} 1.3005(Table3.1) Anal. Calcd for $C_8H_8ClO_4P$: C, 40.94; H, 3.41; Cl, 15,13; P, 13.20. Found C, 40.86; H 3.32; Cl, 15.10; P, 13.15.

Catechol 1-chlorobutyl Phosphate 65g

On analogy with 10a from DMSO (10.64 g, 136 mmol) and 12h (33.56 g, 136 mmol) catechol 1-chlorobutyl phosphate 65g was obtained: 14 g (53%), b.p. 98-100 ^{0}C (0.02 mm) d_4^{20} 1.2999(Table3.2) Anal. Calcd for $C_{10}H_{12}ClO_4P$ C, 45.71; H, 4.57; Cl, 13.52; P, 11.79.Found C, 45.53; H, 4.46; Cl, 13.48; P, 11.45.

1,2,2,2-Tetrachloroethyl Dichlorophosphate 65l

On analogy with 10a from DMSO (9.2 g, 118 mmol) and crude 10d (33.6 g, 118 mmol) 65l was obtained: 20.5 g (58%), b.p. 106-107 ^{o}C (0.01 mm) (Table3.2).Anal. Calcd for $C_2H_1Cl_6O_2P$: C, 7.97; H, 0.33; Cl, 70.76; P, 10.29. Found: C, 7.89; H, 0.38; Cl, 70.69; P, 10.17.

Di(1-chloroethyl) Chlorophosphate 65d

On analogy with 10a from DMSO (3.12 g, 40 mmol) and crude 11a chlorophosphate 65d was obtained: 4.1 g (42%) b.p. 67-69 % (0.1) (Table3.2) Anal. Calcd for $C_4H_8Cl_3O_3P$: C,19.87; H, 3.31; Cl, 44.10, P, 12.84. Found: C, 19.76; H,3.26; Cl, 43.94; P, 12.85.

Di(1-Chloro-2-Methylpropyl) 2,2,2-Trichloroethyl Phosphate 65n

On analogy with 10a from DSMO (2.34 g, 30 mmol and crude 12d (13.85 g, 30 mmol) phosphate 65n was isolated: 5.32 g (37%), b.p. 131 ^{0}C (0.03 mm) Anal. Calcd.for $C_{10}H_{18}Cl_5O_4P$ C, 29.23; H, 4.38; Cl, 43.24; P, 7.55. Found C, 29.01; H, 4.42; Cl, 43.11; P, 7.44.

1-Chloro-2-Methylpropyl Bis (2,2,2-Trichloroethyl) Phosphate 65h.

On analogy with 10a from DMSO (3.2 g, 40 mmol) and crude 12b (18.1 g, 40 mmol) phosphate 65h was obtained: 7.9 g (43%)(Table3.2). Anal.calcd for $C_8H_{12}Cl_7O_4P$: C, 21.26; H, 2.65; Cl, 55.03; P, 6.87. Found: C, 21.13; H, 2.58; Cl, 54.87; P 6.67.

Di(1-Chloroethyl) 2,2,2-Trichloroethyl Phosphate 65o.

On analogy with 10a from DMSO (6.5 g, 80 mmol) and crude 12e (28.19 g, 80 mmol) phosphate 65o was obtained: 11.62 g (41%), b.p. 119 ^{0}C (0.03) (Table 3.2). Anal. Calcd for $C_6H_{10}Cl_5O_4P$: C, 20.31; H, 2.82; Cl, 50.07; P, 8.74. Found : C, 20.23; H, 2.79; Cl, 49.97; P, 8.67.

5.4.2. With t-BuOCl

1-Chloroethyl Dichlorophosphate 65a

BuOCl (12.17 g, 115 mmol) was added dropwise to CCl_4 (50ml) solution of 10a (20.84 g, 115 mmol) at 0-2 ^{0}C. Removing of volatile products and distillation *in vacuo* gave pure phosphate 65a: 6.13 (27%) b.p. 30 ^{0}C(0.04), d_4^{20} 1.4854 (Table3.2).Anal. Calcd for $C_2H_4Cl_3O_2P$: C, 12.15; H, 2.02; Cl, 53.92; P, 15.69. Found C, 12.01; H, 1.97; Cl, 53.08; P, 15.53.

1-Chlorobutyl Dichlorophosphate 65b

On analogy with 10 a from t-BuOCl(5.1 g, 47 mmol) and 10b (9.84 g, 47 mmol) phosphate 65b was obtained: 2.65 g (25%) b.p. 40 ^{0}C (0.025) (Table3.2).Anal. Calcd for $C_4H_8Cl_3O_2P$: C, 21.29; H, 3.54; Cl, 47.18; P.13.74. Found C, 21.19, H, 3.44, Cl, 46.96; P, 13.67.

1-Chloro-2-Methylpropyl Dimethyl Phosphate 65i

Diethyl ether solution of 10c (9.76 g, 16.6 mmol) was dropped into the mixture of methanol (2.98 g, 9.3 mmol) and N,N-diethylalaniline (13.91 g, 93 mmol) in diethyl ether (100 ml) at -10 °C. Reaction mixture was stirred for 2.5 hrs at this temperature. Amine salt was filtered and t-BuOCl (5.06 g, 16.6 mmol) was added dropwise at -5÷-10 °C. Distillation *in vacuo* gave phosphate 65i: (5.38 g (53%), b.p. 48-50 °C (0.04 mm), d_4^{20} 1.1950 (Table 3.2). Anal. Calcd for $C_6H_{14}O_4ClP$: C, 33.25; H, 6.47; Cl, 16.40; P, 14.30. Found C, 33.05; H, 6.34; Cl, 15.98; P, 13.93.

1-Chloro-2-Methylprophyl Diethyl Phosphate 65j

On anology with 65i from ethanol (3.64 g, 79 mmol) , N,N-diethylaniline (11.82 g, 79 mmol), BuOCl (4.5 g, 39.6 mmol) and 10c (8.3 g, 39.6 mmol) was obtained phosphate 65j: 4.64 g (48%), b.p. 75-77 °C (0.04 mm), d_4^{20} 1.1534 (Table3.2).Anal. Calcd for $C_8H_{18}O_4ClP$: C, 39.26; H, 7.36; Cl, 14.49; P, 12.66. Found C, 38.96; H, 7.27; Cl, 14.18; P, 12.13.

Di(1-Chloroethyl) Methyl Phosphate 65m

Di(1-chloroethyl) chlorophosphate 65d (6.25 g, 27 mmol) was dropped into diethyl ether solution of methanol (0.86 g, 27 mmol) and triethylamine (2.73 g, 27 mmol) at 5 °C. The mixture was stirred for 2 hrs at room temperature. Amine salt was filtered. Distillation *in vacuo* gave phosphate 65m: 3.1 g (50%), b.p. 83-84 ^{0}C(0.4 mm). (Table 3.2).Anal. Calcd for $C_5H_{11}Cl_2O_4P$: C, 25.31; H, 4.64; Cl, 29.96; P, 13.08. Found C, 25.31; H, 4.60; Cl, 29.81; P, 13.18.

1-Chloroethyl Dimethyl Phosphate 65k

On analogy with 65m from methanol (1.51 g, 47 mmol), triethylamine (4.8 g, 47 mmol) and 1-chloroethyl dichlorophosphate 65a(4.68g, 21mmol) phosphate 65k was obtained: 1.58 g (35%), b.p. 73-75 ^{0} C(10 mm). (Table3.2).Anal. Calcd for $C_4H_{11}ClO_4P$: C, 25.33; H, 5.80; Cl, 18.73; P, 16.36. Found C, 25.04; H, 5.54; Cl, 18.62; P, 16.05.

5.5. With PCl_5 and Chlorine

A) Cathechol 1-chloro-2-methylpropyl phosphite 12f (15.61g, 63mmol) was added dropwise into CH_2Cl_2 (50ml) solution of PCl_5 (13.12 g, 63 mmol). Distillation *in vacuo* gave *catechol chlorophosphate*66: 8.1 g (70%), b.p. 82-83 °C (0.03 mm); ^{1}H NMR (δ , ppm): 6.8m, C_6H_4; ^{31}P NMR (δ , ppm): 19.

B) A benzene (120 ml) solution of PCl_5 (39.2 g, 187 mmol) was dropped into benzene (15 ml) solution of 12f (57.8 g, 234 mmol). Distillation *in vacuo* gave three fractions: first one is *catechol chlorophosphate*66 (δp 19ppm); second and third fractions are the mixtures of 66 and 65e (3:1 and 1:4, respectelely). Distillation of third fraction gave *cathechol 1-chloro-2-methylpropyl phosphate*65e: b.p. 104 °C (0.02 mm), d_4^{20} 1.3027 (Table3.2). Anal. Calcd for $C_{10}H_{12}O_4ClP$: C, 45.71; H, 4.57; Cl, 13.50; P, 11.79. Found C, 45.23; H, 4.33; Cl, 13.30; P, 11.90.

C) A benzene (15 ml) solution of crude 12a obtained from PCl_3 (12.0 g, 88 mmol), ethanal (17.4 g, 396 mmol) and N,N-diethylaniline (0.656 g, 1.1 mmol) at -1°C was added dropwise into the benzene (60ml) solution of PCl_5 (17.72 g, 88 mmol) at 5°C. The reaction mixture was heated at 35-45°C for 0.5 h. Distillation *in vacuo* gave *di(1-chloroethyl) chlorophosphate*65d:10.1 g (47%), b.p. 61-62 °C (0.05), d_4^{20} 1.4032 (Table3.2). Anal. Calcd for $C_4H_8Cl_3O_3P$: C, 19.87; H, 3.31; Cl, 44.10; P, 12.83. Found C, 19.86; H, 3.13; Cl, 43.86; P, 12.63.

D) Calculated amount of chlorine was passed through the CH_2Cl_2 solution of 12f (48.5 g, 196 mmol) at 10°C. Two resonance signals in ^{31}P NMR spectrum corresponding to P(IV) in phosphates 66 and 65e (10:1) were observed. Distillation *in vacuo* gave *catechol chlorophosphate*66: 27.1 g (73%), b.p. 83-84 °C (0.02 mm). ^{31} P NMR ($\delta_{P,}$ ppm): 19.

E) Calculated amount of chlorine was passed through the toluene (15ml) solution of 12g (30g, 165mmol) at 10 ^{0}C. The reaction mixture was distilled in vacuo to give a fraction: 20g, b.p. 100-118 ^{0}C (0.015 mm) In ^{31}P NMR spectrum two resonance signals at δ19 and 11 ppm were observed (1:3) corresponding to P(IV) in phosphates 66 and 65f. Distillation of this fraction gave two new fractions. In ^{31}P NMR spectra of these fractions ratio of the resonance signals at δp 19 and 11 ppm made up 3:1 and 1:10, respectively.

5.6. Reactions of the Intermediates 11a, 12a and 12b with Aldehydes. Detection of the Secondary Intermediates

A) Ethanal (24.47g, 555mmol) was added dropwise to the mixture of PCl_3 (16.95 g, 123 mmol) and N,N-diethylaniline (0.55 g, 3.7 mmol) at 0. ^{0}C . The mixture was allowed to stand for 24 hrs in a refrigerator. In ^{31}P NMR spectra recorded at 0-5 ^{0}C the resonance signals δp 146, 138 and 163.3, 162.95, 162.43 ppm were observed corresponding to P(III) in 11a and 12a (1.6:1.0). The temperature was allowed to reach 20 ^{0}C and to stand for 2 hrs. After removing of the volatile products *in vacuo* (0.05mm) crude *di(1-chloroethyl) 1-(1-chloroethoxy)ethyl phosphonate* was obtained: 18.8g (65%) (Table 3.1), $d^{20}{}_4$.1.3135. Anal. Calcd C, 30.62; H, 5.10; Cl, 33.97; P, 9.89. Found 30.12; H, 5.01; Cl, 34.12; P, 9,80.

B) *1-(1-chloroethoxy)ethyl phosphorodichloridite*28a. Acetaldehyde (3.92 g, 89 mmol) was dropped to the mixture of PCl_3 (12.52 g, 89 mmol) and N,N-diethylaniline (0.106 g, 0.7 mmol) at -25 ^{0}C. The reaction mixture was stirred for 25 hrs at this temperature. ^{31} P NMR(δ, ppm): 219(PCl_3); 178.1, 178.8 10a; 169.6, 169.3 28a (22:1:7). ^{1}H NMR (δ, ppm): 1.56 d, J_{HH} 5.2 Hz, 3H, $O_2CH\underline{CH_3}$; 1,73 d, J_{HH} 5.5Hz, 3H, $OCHCl\underline{CH_3}$. 5.75 k, J_{HH} 5.5Hz, 1H, OCHCl; 5.92 dk, J_{HH} 5.2Hz, J_{PH} 12.5Hz (28a, intensive signals); 1.31d, J_{HH} 5.1Hz, CH_3; δ 4.98 k, J_{HH} 5.1Hz, CH (paraldehyde, weak signals); 1.82 d J_{HH} 5.6 Hz, CH_3; 6.3 dk J_{HH} 5.6 Hz, J_{PH} 13.9 Hz (10a, weak signals); ^{13}CNMR (δ, ppm): 22.68 s $O_2CH\underline{C}H_3$, 25.90 s $OCHCl\underline{C}H_3$, 88.67 s, 88.61 s, OCHCl, 98.64 d, 98.47 d, J_{PC} 4.4Hz, 98.13 d, J_{PC} 5.5Hz POCH (28a, very intensive signals); 20.00 s, CH_3, 97.60 s CH(paraldehyde, weak signals), 25.50 s CH_3, 86.21 d J_{PH} 10.6Hz, CH(10a, weak signals). The temperature was allowed to reach -10^{0}C. Exothermic effect was observed and temperature reached 25^{0}C. The volatile products were collected in a trap cooled by liquid N_2*in vacuo*. Distillation of volatile products using Vigreux column gave *1-chloro-1-(1-chloroethoxy)ethane*16a with a small admixture of papaldehyde : 4.11 g (62%), b. p. 114-115^{0}C, d_4^{20} 1.1300 (113-114^{0}C, d_4^{20}, 1.1320 [73]).^{1}H NMR (δ , ppm): 1.1 d J_{HH} 6,3 Hz, 6H, CH_3; 5.71 k J_{HH} 6.3 Hz, 1H, CH.

C) *Reaction of secondary intermediate*28a(R=Pr) *with chlorine*. Butanal (19.85 g, 275 mmol) was added dropwise into the mixture of PCl_3 (18.9 g, 138 mmol) and N,N-diethylaniline (0.154 g, 1 mmol) at -20 ^{0}C. The reaction mixture was allowed to stand for 1.5 hrs at this temperature. A calculated amount of chlorine was passed through the reaction mixture. Distillation afforded pure $POCl_3$: 17.13 g , b.p. 43-44 ^{0}C (73 mm) , δp 0.0 ppm and *1-chloro-1-(1-chlorobutoxy) butane*16c : 12.72 g, b.p. 76-79 ^{0}C (16 mm), $d^{20}{}_4$ 1.0297 (b.p.75-78 ^{0}C (16 mm), $d^{20}{}_4$ 1.0330 [73]). Anal. Calcd for $C_8H_{16}Cl_2O$: C, 48.26; H, 8.10; Cl, 35.61. Found C, 48.06; H, 7.94; Cl, 35.47.

D) *Reaction of secondary intermediate*28a(R=i-Pr)*with chlorine*. On analogy with synthesis of 16c from PCl_3 (14.38 g, 105 mmol), 2-methylpropanal (15.14 g, 210 mmol) and N,N-diethylaniline (0.118 g, 0.08 mmol) $POCl_3$: 8g , b.p. 44-45 ^{0}C (80 mm) and *1-chloro-2-methyl-1-(1-chloro-2-methylpropoxy)propane*16b : 9.49 g , b.p. 75-78^{0}C (18 mm), $d^{20}{}_4$ 1.0658; were obtained. ^{1}H NMR(δ, ppn): 1.05 d, J_{HH} 6.3 Hz, 12H, CH_3; 2.17 m 2 H <u>CH</u>Me_2; 5.70 d, J_{HH} 4.0 Hz, 2H, CHCl. Anal. Calcd for $C_8H_{16}Cl_2O$: C, 48.26; H, 8.10; Cl, 35.61. Found C, 48.11; H, 7.91; Cl, 35.38.

Reaction of Bis (2,2,2-trichloroethyl) Phosphorochloridite 46a with 2-methypropanal in the Presence of N,N-diethylaniline (1:3:0.002)

2-Methylpropanal (9.59 g, 132 mmol) was dropped into the mixture of 46a (16.1 g, 0.44 mmol) and N.N-diethylaniline(0.13 g, 0.087 mmol) at -4-0 ^{0}C. The temperature was allowed to reach 25 C for 0.5hr. The ^{31}P NMR spectra were recorded (Section 4). After 24hrs standing at room temperature reaction mixture was distilled *in vacuo* and elimination of HCl was observed.

Bis(2,2,2-trichloroethyl) 2-methyl-1-(2-methylpropen-1-yloxy)propylphosphonate 26c was isolated: 8.6 g (29.9%), b.p. 185-186 ^{0}C (0.04 mm), $d^{20}{}_4$ 1.3635 (Table 3.1).Anal. Calcd for $C_{12}H_{19}Cl_6O_4P$: C, 30.57; H, 4.03; Cl, 45.22; P, 6.58. Found C, 30.15; H, 4.11; Cl, 45.10; P, 6.23.

^{1}H (300MHz), ^{13}C (75.5 MHz) and ^{31}P (121.5 MHz) NMR spectra were recorded on a 7.0 Tesla IBM/Brucker AF 300 NMR spectrometer in $CDCl_3$ and are reported in δ units with Me_4Si as internal or 85% H_3PO_4 as external standard. The spectra are fairly complicated, because virtually every compound is a mixture of stereo isomers from the chiral RCHClO-group and in some cases a chiral center at phosphorus. No attempt has been made to sort out the separate spectra of these isomers. Reagents and solvents were purified by using standard procedures. All experiments were conducted under dry nitrogen.

REFERENCES

[1] Gazizov, M.B.; Khairullin R.A.; Moskva V.V. *Usp. Khim.* 1990, 59, 431-456

[2] Gazizov, M.B.; Khairullin, R.A. Itogi nauki i techniki. VINITI. *Ser. Org. Khim.* 1990, 15.1-94.

[3] Fossek, W. *Monatsh.*, 1884, 5, 121-127.

[4] Michaelis, A. *Chem. Ber.*1885, 18, 898-910.

[5] Conant, J.B.; Mc.Donald, A.D. *J. Amer. Chem. Soc.* 1920, 12, 2337-2348.

[6] Conant, J.B.; Vallingford, V.H. *J. Amer. Chem. Soc.* 1924, 46, 192-203

[7] Page, H.J. *J. Chem. Soc.* 1912, 101, 423-431.

[8] Kabachnik, M.I. *Russ. Chem. Rev.* 1947, 16. 402-420

[9] Kabachnik, M.I.; Shelepova, S.S. Dokl, Akad. *Nauk SSSR* 1950, 75, 219-222.

[10] Kabachnik, M.I.; Shelepova, S.S. Jzvest. Akad. *Nauk SSSR* 1950, 39-46.

[11] Atherton, F.R,; Clark, V.M,;Todd, A.R. *Rec.Trav. Chim.* 1950, 69, 295-299

[12] Pudovik, A.N.; Khairullin, V.K. *Usp. Khim.* 1968, 37, 745-785.

[13] Aliev, P.S.; Khairullin, V.K.. Zhur. *Obshch. Khim.* 1976, 46,266-272.

[14] Novikova, Z.S.; Odinets, J.L.; Lutsenko, J.V. Zhur. *Obshch. Khim.* 1983,53, 1673-1678.

[15] Golovanov, A.V.; Maslennikov, I.G.; Gudina, J. V. Zhur. *Obshch. Khim.* 1985, 55, 2623-2627.
[16] Razumov, A.I.; Gurevich, P.A.; Nurtdinov, S. Kh., Muslimov, S.A.. Zhur. *Obshch. Khim.* 1977, 47, 1421-1425.
[17] Bagrov, F.V.; Vasoleva, L.A.; Stifanova, Ju.P., Kuznetsova, O.S. Zhur. *Obshch. Khim.* 1979, 49, 1242-1246.
[18] Golovanov,A.V.; Maslennikov, I.G.; Lavrentev, A.N. Zhur. *Obshch. Khim.* 1986, 56, 1957-1960.
[19] Golovanov, A.V.; Maslennikov, J.G.; Lebedev, V.B. Zhur. *Obshch. Khim.* 1986, 56, 2540-2542.
[20] Miller, J.A.; Nunn, M.J. *Tetrahedron Lett* 1972, 38, 3953-3956.
[21] Miller, J.A.; Nunn, M.J. *J. Chem. Soc. Perkin* I 1976, 535-539.
[22] Nurtdinov, S.Kh.; Savran, V.I.; Zikova, T.V.; Salachutdinov, R.A.; Tsivunin, V.S. Zhur. *Obshch. Khim.* 1979, 49, 2446-2452.
[23] Michil, J.K.; Miller, J.A.; Nunn, M.J.; Stewart, D. J. *Chem. Soc. Perkin* I 1981, 1744-1749.
[24] De Ath, N.J.; Miller, J.A.; Nunn, M.J.; Stewart, D. *J.Chem.Soc. Perkin* I 1981, 776-787.
[25] Shermolovich, Ju.G.; Solobev, A.V.; Danchenko, E.A.; Markovski, L.N. Zhur. *Obshch. Khim.* 1985, 55, 291-297.
[26] Nurtdinov, S.Kh.; Sultanova, R.B.; Mukhametova, G.G. Zhur. *Obshch. Khim.* 1989, 59, 344-347.
[27] Levin, J.A., Gozman, E.E. Jzvest. Akad. *Nauk SSSR.* 1970, 173-175.
[28] Gazizov, M.B.; Razumov, A.I. Zhur. *Obshch. Khim.* 1969, 39, 2600-2602
[29] Gazizov, M.B.; Sobchuk, T.I.; Razumov, A.I. Zhur. *Obshch. Khim.*1969, 39, 2595.
[30] Gazizov, M.B.; Sultanova, R.B.; Moskva, V.V. Zhur. *Obshch. Khim.*1971, 41, 932-933.
[31] Gazizov, M.B.; Sultanova, R.B.; Razumov, A.I.; Ostanina, L.P. Zhur. *Obshch. Khim.* 1973, 73, P.213-214.
[32] Gazizov, M.B.; Sultanova, R.B.; Razumov, A.J.; Ostanina, L.P.; Zikova, N.V.; Salachutdinov, R.A. Zhur. *Obshch. Khim.* 1973,43, 2160-2165.
[33] Gazizov, M.B.; Zikova, N.V.; Anoshina, N.P.; Sultanova, R.B.; Salachutdinov, R.A. Zhur. *Obshch. Khim.* 1978,48, 1979-1984.
[34] Gazizov, M.B.; Sultanova, R.B.; Razumov, A.J.; Zikova, N.V.; Pachinkin, A.P.; Salachutdinov, R.A. Zhur. *Obshch. Khim.* 1976,46, 1223-1228.
[35] Tsivunin, V.S.; Afanasev, J.N.; Ivanova, R.G. Zhur. *Obshch. Khim.* 1968, 38, 1523-1527
[36] Gazlzov, T.Kh. Arbuzov Reaction.Kazan: *Science*, 2003 1-319(in Russian).
[37] Kirby, A.J.; Warren, S.G..The Organic Chemistry of Phosphorus. Moscow: Mir, 1971,1-403(in Russian).
[38] Gazizov, M.B.; Kachalova, T.N.; Karimova, R.F.; Khairullin, R.A.; Sinyashin, O.G. Zhur. *Obshch. Khim.* 1997, 67, 2055-2056.
[39] Gazizov, M.B.; Kachalova, T.N.; Pudovik, M.A.; Khairullin, R.A.; Sinyachin, O.G.; Cadikova, A.I.; Gazizov, K.M.; Musin, R.S. Zhur. *Obshch. Khim.* 2001, 71, 523-524.
[40] Nurtdinov, S.Kh.; Savran, V.J.; Zikova, T.V.; Salachutdinov, R.A.; Tsivunin, V.S. Leningrad.: LTI 1980, 88-95.
[41] Nurtdinov, S.Kh.; Sultanova, R.V.; Mukhametova, G.G.; Zikova, T.V.; Shagvaleev, F.Sh. Zhur. *Obshch. Khim.* 1989, 5, 344-347.

[42] Plets, V.M. Organicheskie soedinenija phosphora. M: *Oborongiz* 1940, 139-140.
[43] Gazizov, M.B.; Khairullin, R.A.; Kadirova, R.F.; Lewis, E.S.; Kook, A.M. *J.Chem. Soc., Chec. Chommun.* 1990,16,1133-1134.
[44] Gazizov, M.B.; Levis, E.S.; Kook, A.M.; Khairullin, R.A.; Kadirova. R.F. Zhur. *Obshch. Khim.* 1990, 60,1489-1191.
[45] Gefte,r E.L.; Rogacheva, J.A. Zhur. *Obshch. Khim.* 1962, 32, 3952-3965.
[46] Pudovik, A.N.; Faizullin, E.M. Dokl, Akad. *Nauk SSSR.* 1965, 165, 586-588.
[47] Pudovik, A.N.; Faizullin, E.M. . Zhur. *Obshch. Khim.* 1966, 36, 798-801.
[48] Gazizov, T.Kh.; Pashinkin, A.P.; Kharlamov, V.A. Zhur. *Obshch. Khim.* 1977, 47,1226-1230.
[49] Gazizov, M.B.; Khairullin, R.A. VIII Conference on Chemistry and Application of Organophosphorus Compounds: Abstracts of papers. Kazan (USSR), 1985, Part I, P. 89.
[50] Gazizov, M.B.; Shergina, i.I.; Khairullin R.A.. Zhur. *Obshch. Khim.* 1984, 54,1204-1205.
[51] Gazizov, M.B.; Khairullin, R.A.; Ostanina, L.P. International Conference on Phosphorus Chemistry Abstracts of papers, 30, Bonn, (BRD) 1986. A.30.
[52] Gazizov, M.B., Khairullin, R.A. Zhur. *Obshch. Khim.* 1986, 56, 226-227.
[53] Birum, G: Patent USA 3014994(1961), *Chem. Abstr.* 1962, 56,6, 11622f.
[54] Gazizov, M.B.; Zakharov, V.V.; Khairullin, R.A.; Moskva, V.V. Zhur. *Obshch. Khim.* 1984, 51, 671 and 2783.
[55] Gefter, E.L.; Kabachnik, M.I. Dokl, Akad. *Nauk SSSR.* 1957, 114, 541-543.
[56] Gazizov, M.B.; Khairullin, R.A.; Moskva, V.V.; Saveleva, E.I.; Ostanina, L.P.; Nikolaeva, V.G. Zhur. *Obshch. Khim.* 1990, 60, 1766-1775.
[57] Gazizov, M.B.; Khairullin, R.A.; Kadirova, R.F. Zhur. *Obshch. Khim.* 1989, 59, 1757-1459
[58] Gazizov, M.B.; Khairullin R.A.; Shergina I.I. Zhur. *Obshch. Khim.* 1984, 54, 2638.
[59] Nifantev, E.E.; Vasyanina, L.K. P^{31} NMR spectroscopy; Moscow, 1986, 1-148.
[60] Mark, V.; Dungan, C.H.; Crutchfield, M.M.; Van-Wazer, J.R. P^{31} Nuc lear Magnetic Resonance. John Wiley and Sons, Inc. USA, 1967; Vol. 5, Chapter 4, 227-457.
[61] Gazizov, M.B.;Razumov, A.I.; Khairullin, R.A. Zhur. *Obshch. Khim.*1980, 50, 231-232.
[62] Pudovik, A.N.; Konovalova, J.V.; Ishmaeva, E.A. Reaktzii i metodi issledovanija organicheskich soedinenii; Khimia, Moskva, 1973, Book 23, 1-488.
[63] Nurtdinov, S.Kh.; Kazarin, S.V.; Ismagilova, N.M.; Tsivunin V.S. Zhur. *Obshch. Khim.*1976, 46, 36-39.
[64] Gazizov, M.B.; Sultanova, R.B.; Ostanina, L.P. Zhur. *Obshch. Khim.*1971, 41, 2167-2171.
[65] Grofts, P.C.; Markes, J.N.; Pydon, H.N. *J. Chem. Soc.* 1958, 4250-4254.
[66] Gazizov, M.B.; Khairullin, R.A.; Karimova, R.F. Zhur. *Obshch. Khim.*, 1995, 65, 1227-1228
[67] Lutsenko, I.F; Kraits, Z.S.; Proskurnina, M.V. Dokl, Akad. *Nauk SSSR.* 1963, 148, 846-849.
[68] Kim, T.V.; Ivanova, Zh.M.; Gololobov, Ju.G. Zhur. *Obshch. Khim.* 1978, 48, 346-353.
[69] Cloede, J.; Nikolajezek, M.; Lopusinski, A. *J. fur pract. Chem.* 1974, 316, 703-704.

[70] Gazizov, M.B.; Zakharov, V.V.; Khairullin, R.A.; Moskva, V.V. Zhur. *Obshch. Khim.* 1986, 56, 1659-1660.
[71] Timokhin, B.V.; Kazantseva, M.V.; Donskich, V.J. Zhur. Obshch. Khim. 1989, 59, 2486-2492.
[72] Petrov, K.A.; Chauzov, V.A.; Agafonov, S.V. *Usp. Khim.* 1982, 52, 412-437.
[73] Pokonova, J.V. Chemistry and Technology α-Halogenoethers. Leningrad: Leningrad University, USSR, 1982, 1-271.

In: New Developments in Aldehydes Research
Editors: L. Torrioni and E. Pescasseroli

ISBN: 978-1-62417-090-4

Chapter 2

SYNTHESIS OF HETEROCYCLIC COMPOUNDS BY INTERACTION OF ALDEHYDES WITH MONOTERPENOIDS

Oksana S. Mikhalchenko, Konstantin P. Volcho* and Nariman F. Salakhutdinov

N.N.Vorozhtsov Novosibirsk Institute of Organic Chemistry,
Siberian Branch, Russian Academy of Sciences,
Novosibirsk, Russian Federation

ABSTRACT

The review covers the reactions of aldehydes with monoterpenoids leading to chiral oxygen-containing heterocyclic compounds of various structural types. Almost all these reactions are catalyzed by acids, the Lewis acids and montmorillonite clays being typically used as the catalysts. Some of the resulting compounds exhibited significant biological activity. The most complex multistage transformations are observed when performing the reactions of aldehydes with oxygen-containing *para*-menthane, pinane and carane monoterpenoids in the presence of clays. Due to the ability of monoterpenoids of different structural types to rearrange into carbocations with the *para*-menthane framework upon protonation, a number of identical heterocyclic compounds can be obtained using various starting terpenoids.

INTRODUCTION

Reactions of aliphatic and aromatic aldehydes with monoterpenoids in the presence of acid catalysts often lead to new chiral oxygen-containing heterocyclic compounds with different types of framework. Various types of monoterpenoids that are widely distributed in nature can be involved in these reactions. The reactions typically proceed with high

* Fax: +7 383 3309752; e-mail: volcho@nioch.nsc.ru.

stereoselectivity. Montmorillonite clays and Lewis acids are usually used as the catalysts for these transformations. The reaction routes strongly depend on terpenoid structure and a type of the catalyst used.

Some of the products possess pronounced biological activity. Thus, it is a new and promising way for one-pot synthesis of new types of pharmacologically active compounds from available and inexpensive aldehydes and monoterpenoids.

Despite a variety of attractive features, the use of aldehydes in reactions with monoterpenoids to synthesize heterocyclic compounds began as late as fifteen years ago, however, research in the field is being intensively developed now. To this moment, only a few reviews partly concerning the field have been published [1-3].

The review consists of three parts: reactions of aldehydes with acyclic, monocyclic and bicyclic monoterpenoids. It covers the reactions of aldehydes with monoterpenoids of various structural types published in literature up to the beginning of 2012. The possible mechanisms of almost all transformations are briefly discussed in the review.

Interaction of Aldehydes with Acyclic Monoterpenoids

Acyclic monoterpenes and their oxygen-containing derivatives are abundant in nature, some of them are produced at the industrial scale. Thus, myrcene 1 is a valuable starting compound for the synthesis of fragrance substances.

The interaction between myrcene 1 and (2*E*)-butenal 2 in an autoclave at 200 °C is known to yield the mixture of Diels–Alder adducts 3 and 4 at a 2:3 ratio [4] (Scheme 1).

It was ascertained that if montmorillonite clay was used as a catalyst for this reaction, the interaction occurred at carbonyl group of aldehyde 2 instead of the olefinic double bond and yielded a heterocyclic compound 5 with the 3-oxabicyclo[3.3.1]non-6-en framework as the main product [5, 6].

The presumable mechanism of this reaction includes the formation of an intermediate cation with the *para*-menthane framework 6, which subsequently interacts with the aldehyde that plays the role of a nucleophile [5, 6].

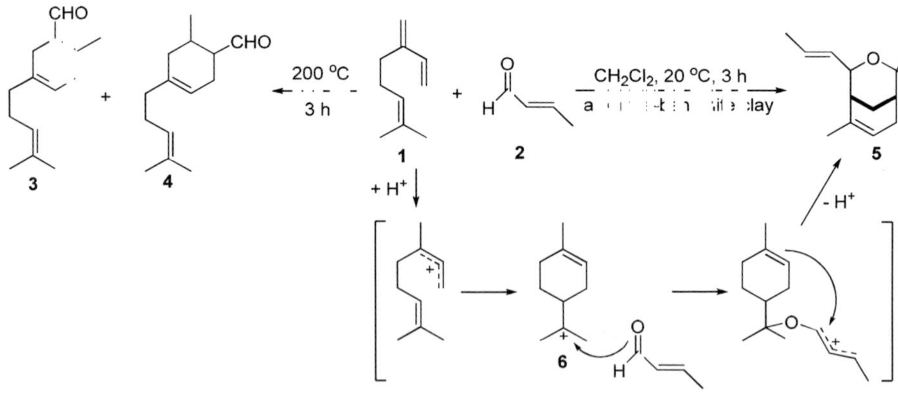

Scheme 1.

Geraniol 7 in the presence of a Lewis acid $BF_3 \cdot Et_2O$ interacts with various aldehydes 8–25 forming 3-oxabicyclo[3.3.1]nonenes 27-44, which are the analogues of compound 5, and substituted tetrahydropyrans 26a,b (Scheme 2, Table 1) [7].

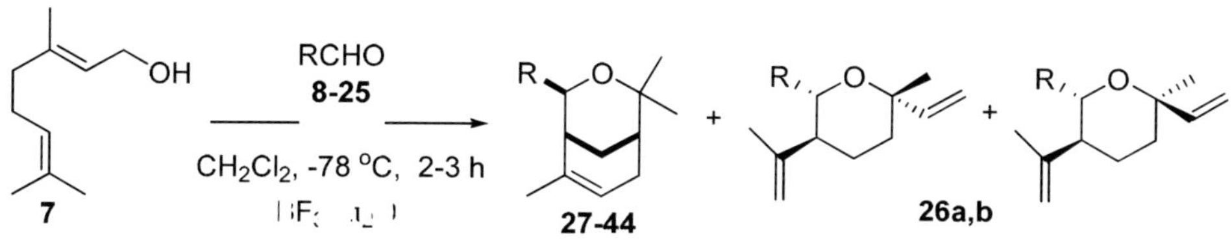

Scheme 2.

The reaction obviously proceeds via the protonation of the hydroxyl group of alcohol 7 followed by eliminating of the water molecule, yielding cation 45, and cyclization (Scheme 3, pathway *a*) yielding the previously mentioned carbocation 6 with the *para*-menthane framework [7]. This cation can subsequently interact with an aldehyde molecule, yielding the oxocarbenium ion 46, in which oxonium-ene cyclization results in the formation of the oxabicyclic products (27-44).

Table 1. Reactions of geraniol 7 with aldehydes 8-25 [7]

No	Aldehyde R =	Products yield, % (No)		No	Aldehyde R =	Products yield, % (No)	
			26a,b				26a,b
8	C_6H_5-	52 (27)	25	17	4-MeO_2C-C_6H_4-	57 (36)	0
9	4-Cl-C_6H_4-	70 (28)	0	18	4-OH-C_6H_4-	60 (37)	0
10	4-Br-C_6H_4-	70 (29)	0	19	4-CH_3-C_6H_4-	68 (38)	0
11	4-F-C_6H_4-	68 (30)	0	20	Cyclo-C_6H_{11}-	62 (39)	0
12	4-NO_2-C_6H_4-	50 (31)	15	21	C_2H_5-	8 (40)	65
13	2-NO_2-C_6H_4-	54 (32)	0	22	C_3H_7-	8 (41)	65
14	2-Naphthyl-	55 (33)	0	23	$(CH_3)_2CHCH_2$-	10 (42)	67
15	C_6H_5-CH=CH-	62 (34)	0	24	$CH_3(CH_2)_5$-	15 (43)	60
16	4-NO_2-C_6H_4-CH=CH-	55 (35)	0	25	$CH_3(CH_2)_{14}$-	10 (44)	62

The formation of compounds 26a,b may proceed via the interaction between the aldehyde and acyclic cation 45 (Scheme 3, pathway *b*) followed by hetero- and carbocyclizations [7]. The formation of isomer 26a is more favorable for steric reasons, it actually is observed regardless of structure of the aldehydes used.

The interaction between geraniol 7 and aromatic aldehydes 8-19 or cyclohexanecarboxaldehyde 20 in the presence of $BF_3 \cdot Et_2O$ results in formation of compounds 27-39 as the major products with yields of 50–70% [7]. Meanwhile, the interaction with aliphatic aldehydes (propionaldehyde 21, butyraldehyde 22, 3-methylbutanal 23, heptanal 24 and palmitaldehyde 25) mostly results in the formation of tetrahydropyrans 26a,b with yields of 60–65%. This can presumably be attributed to the fact that aliphatic aldehydes possess higher nucleophility and smaller size compared to aromatic aldehydes, which facilitates their interaction with the carbocation 45.

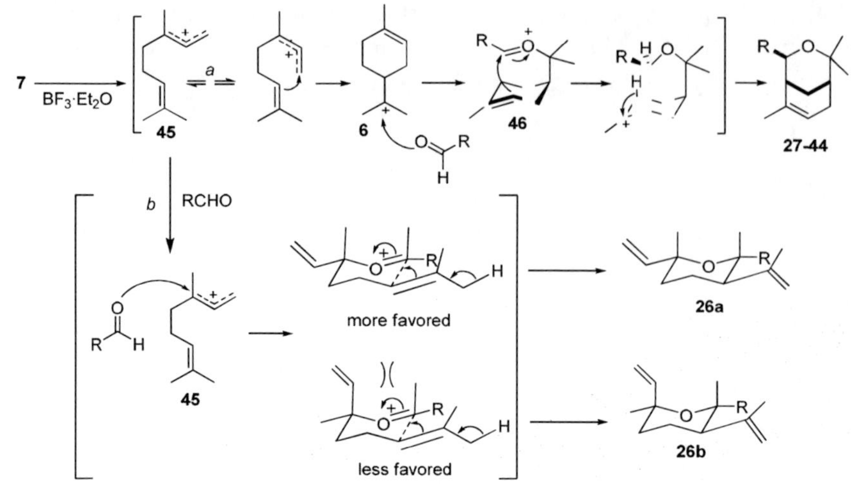

Scheme 3.

The formation of a noticeable amount of tetrahydropyran products 26a,b in the case of benzaldehyde 8 (25%) and 4-nitrobenzaldehyde 12 (15%) can be attributed to the electron-withdrawing effect of phenyl and *para*-nitrophenyl groups, which destabilize the oxocarbenium ion 46. On the contrary, the presence of electron-donating groups (-OH, $-CH_3$) in the fourth position of the aromatic ring stabilizes this ion [8].

The interaction between geraniol 7 and salicylaldehyde 47, which results in the formation of tetracyclic compound 48 with yield of 40%, should be considered separately. The formation of carbocation 49 is presumably followed by nucleophilic attack of the hydroxyl group at cation center and by the formation of compound 48 (Scheme 4) [7].

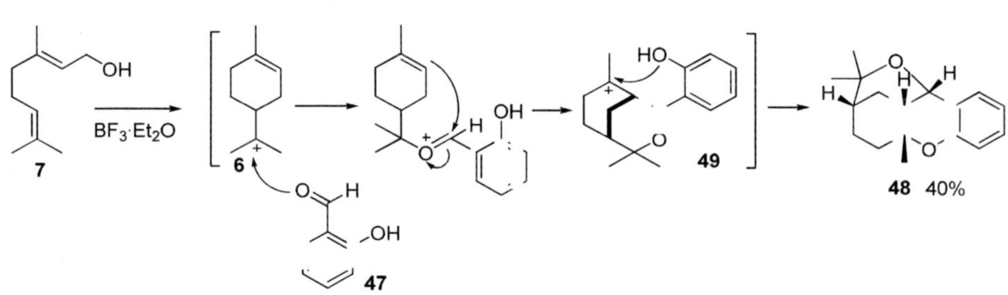

Scheme 4.

Geranial 50 and neral 51 are unsaturated aldehydes and participate in homocoupling reactions catalyzed by bis(cyclopentadienyl)titanium (III) chloride in the presence of water. The reactions yield lactols 52 and 53, respectively, which are formed as diastereomers with respect to the C-2 position (Scheme 5) [9].

Under identical conditions, geranial 50 participates in the cross-coupling reactions with decanal 54 and 3-phenylpropanal 55 to yield the mixture of isomers 56 and 58, 57 and 59, respectively, *E*-isomers 56 and 57 being predominant [9] (Scheme 6). The mixtures of the same products are formed upon interaction between neral 51 and aldehydes 54 and 55, isomers 58 and 59 being predominant.

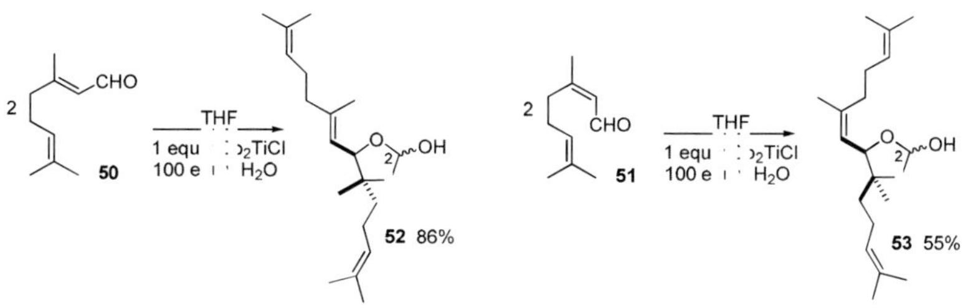

Scheme 5.

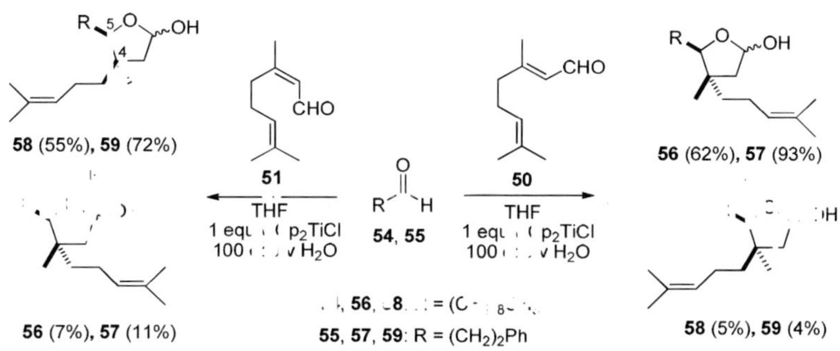

Scheme 6.

The authors of [9] have assumed that aquacomplex 60 is formed in the presence of water (Scheme 7). The inner-sphere single-electron transfer from Ti^{III} as a result of coordination between the aquacomplex 60 and conjugated alkenal 50 or 51 would give a titanoxy-allyl-type radical 61 (Scheme 7). In this titanoxy derivative the Ti^{IV} atom has only 16 electrons in the valent shell and therefore can coordinate again with the oxygen atom of aldehyde 54 to give an intermediate 62. This intermediate shows the suitable spatial arrangement to facilitate the required overlap between the π-orbitals of the carbonyl group and the delocalized allylic radical. Thus, the crucial coupling step from 23 to 24 might be viewed as a 7-*endo-dig* cyclization favored by Baldwin's rules [10]. Moreover, the stereoselective formation of 56 from the *E*-alkenal 50 and 58 from the *Z*-alkenal 51 suggests retention of the original alkene configuration in the delocalized allyl-type radicals 62 and 63 [9].

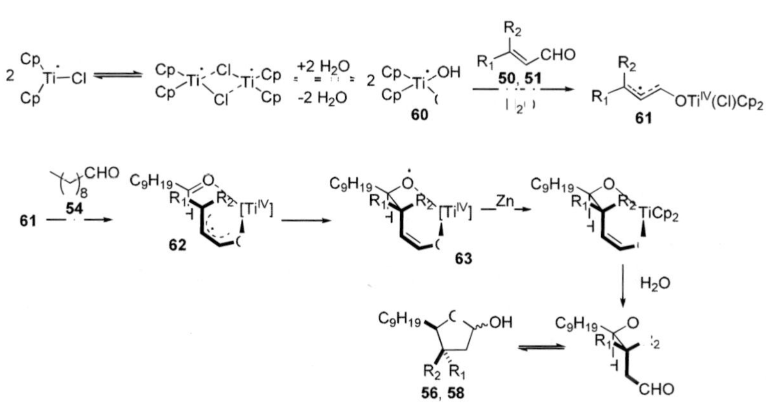

Scheme 7.

Citral 64, which is the mixture of geranial 50 and neral 51, can interact with salicylaldehyde 47, 5-chloro-2-hydroxybenzaldehyde 65, and 5-bromo-2-hydroxybenzaldehyde 66 in the presence of Na_2CO_3, yielding chromene-3-aldehydes 67-69 (Scheme 8) [11, 12]. Let us mention that compounds 67-69 were subsequently used in the synthesis of polycyclic chromene derivatives, which exhibit high cytotoxic activity [12].

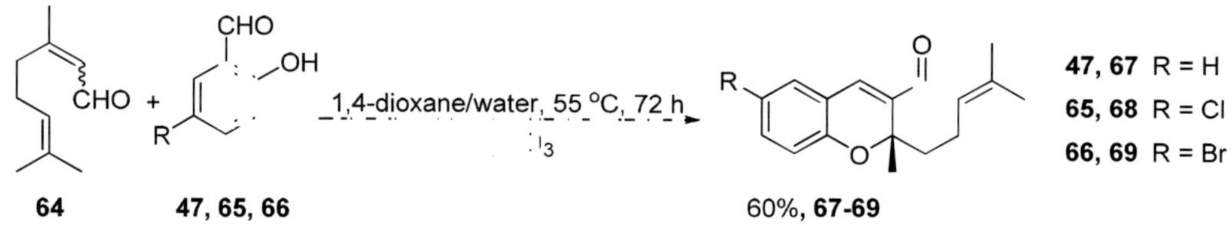

Scheme 8.

The Lewis and Brønsted acids are known to catalyze the reactions between homoallylic alcohols and aldehydes yielding a large number of tetrahydropyran derivatives [13–18]. (*R*)-Citronellal 70 undergoes intramolecular carbonyl-ene cyclization under acidic conditions, yielding the mixture of diastereomeric secondary alcohols with the *para*-menthane framework, the ratio between the alcohols depending on the nature of the acid and reaction conditions. Homoallylic alcohol isopulegol 71, which can subsequently interact with aldehydes, is usually the main reaction product [19].

Indeed, in the presence of aldehydes and catalytic amounts of $Sc(OTf)_3$ at -78°C, (*R*)-citronellal (*R*)-70 both undergoes cyclization into isopulegol 71 and interacts with aldehydes, yielding the mixture of diastereomeric products with the chromene framework (Scheme 9) [20].

Investigation of the interaction between citronellal (*R*)-70 and various aromatic (benzaldehyde 8, 4-chlorobenzaldehyde 9, 4-bromobenzaldehyde 10, 4-nitrobenzaldehyde 12, 2-nitrobenzaldehyde 13, 4-methylbenzaldehyde 19, 1-naphtaldehyde 72 and 4-methoxybenzaldehyde 73) and aliphatic (cyclohexanecarboxaldehyde 20, propionaldehyde 21, 3-methylbutyraldehyde 23, 3-phenylpropanal 55) aldehydes has demonstrated that the corresponding chromenes 74-85 are formed in all cases with appreciably high yields (70–90%). The diastereomer a was the main product in all the reactions, the ratio a:b between the diastereomers varied from 8:2 to 9:1. The reaction duration has varied from 10 h (for benzaldehyde 8) to 14 h (in the reaction with 2-nitrobenzaldehyde 13).

It is an interesting fact that *trans*-3-phenyl-2-propenal 15, which is sensitive to acids, also participates in the reaction, giving isomeric products 86a,b at the 9:1 ratio with the yield of 85% [20].

As mentioned previously, the possible mechanism of the reaction comprises the cyclization of citronellal (*R*)-70 into isopulegol 71 and interaction of 71 with the aldehyde, giving carbocation 87 (Scheme 10) [20].

Under conditions of Asinger condensation [21, 22], (*S*)-citronellal (*S*)-70 interacts with 3-hydroxy-2,2-dimethylpropanal 88 and ammonia, acting as an oxo-component, yielding diastereomeric products 89 and 90 at a 2:1 ratio (Scheme 11) [22].

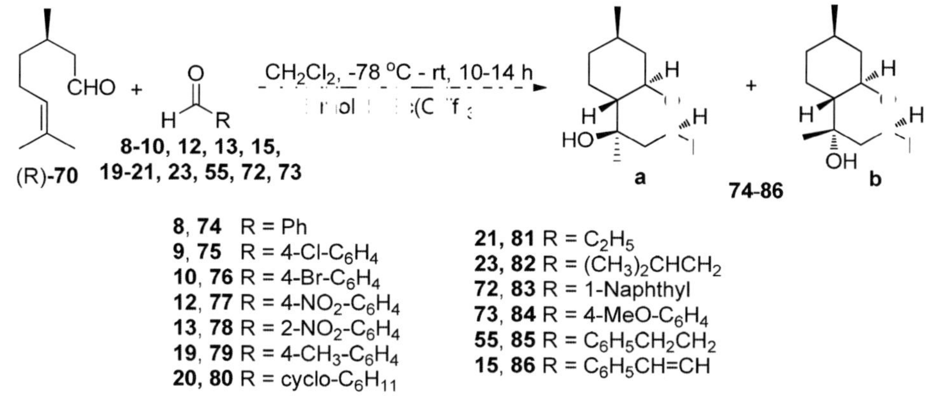

Scheme 9.

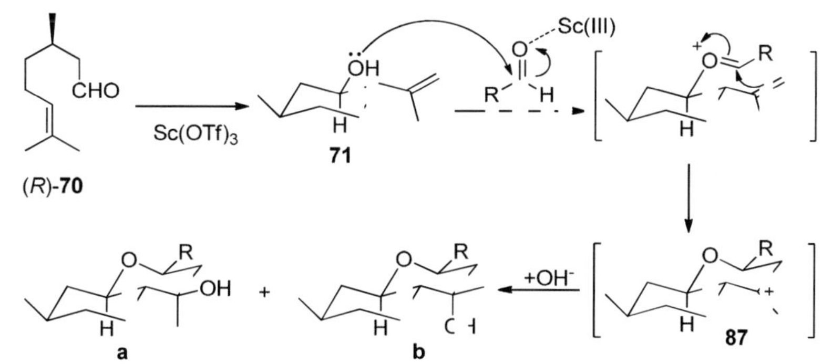

Scheme 10.

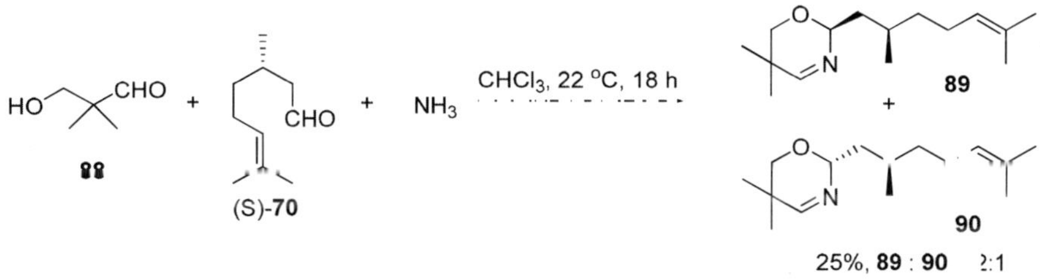

Scheme 11.

Interaction of Aldehydes with Monocyclic Monoterpenoids

The compounds with the *para*-menthane framework are the most abundant natural monocyclic monoterpenoids, (*R*)-limonene (*R*)-91, the main component of citrus essential oil, being a typical representative of this series of monoterpenoids.

Dipentene 91, which is an equimolar mixture of the (*R*)- and (*S*)-limonenes (*R*)-91 and (*S*)-91, interacts in the presence of montmorillonite clay with aromatic (8, 18, 73, 94) and aliphatic ((2*E*)-butenal 2, butyraldehyde 22, acrylaldehyde 92 and methacrylaldehyde 93) aldehydes, giving heterocyclic products (Scheme 12).

The formation of bicyclic compounds with the 3-oxabicyclo[3.3.1]non-6-en framework (Scheme 12) is observed in all cases [6, 23]. As already mentioned, compounds of this type can be obtained via the interaction between geraniol 7 and aldehydes in the presence of $BF_3{\cdot}Et_2O$ (Scheme 2), however, the use of monoterpene 91 as a starting compound makes it possible to obtain bicyclic ethers without the formation of side products 26a,b. Furthermore, the use of the optically active (*R*)- и (*S*)-limonenes 91 in these reactions leads to the formation of optically active heterocyclic compounds. Thus, the reactions between (*R*)- and (*S*)-limonenes 91 and 4-hydroxybenzaldehyde 18 or 4-hydroxy-3-methylbenzaldehyde 95, which are catalyzed by montmorillonite clay, give chiral products 37 and 100, which are antagonists of estrogen receptor α and β [24, 25].

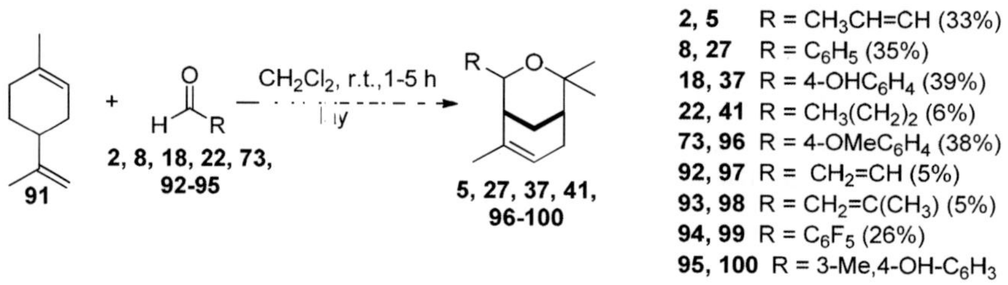

Scheme 12.

(*R*)-Limonene (*R*)-91 and racemic 1,2,4-trimethyl-4-isopropenyl-cyclohexene 101 interact with salicylaldehyde 47, yielding tetracyclic compounds 48 and 102, respectively (Scheme 13). Identically to the transformations with geraniol 7, the reaction is not terminated at the stage where the bicyclic compounds are formed. The cationic center interacts with the hydroxyl group, giving the products containing the xanthene framework condensed with the tetrahydropyran ring [6, 26].

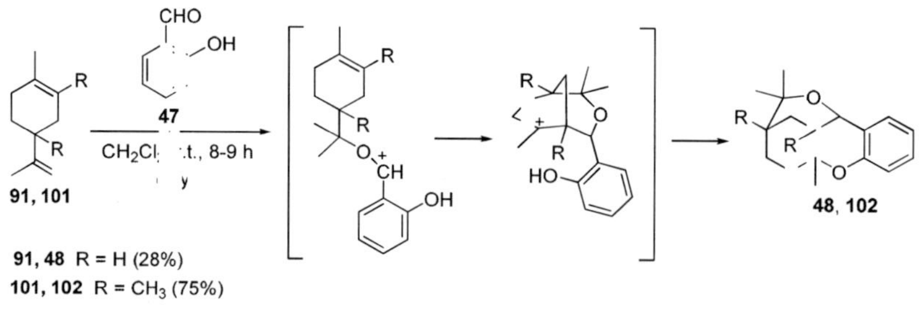

Scheme 13.

If limonene (*R*)-91 is used as the starting compound, the yield of tetracyclic product 48 is small (28%) [6]. It was found [27] that the application of aldehyde 47 onto clay in a solvent followed by distillation of the solvent, and the subsequent application of monoterpene 91 in the solvent followed by the repeated distillation of the solvent allows one to reduce the reaction time from 8 to 0.5 h and increase the yield of compound 48 up to 73%.

Terpinolene 103, which is the isomer of limonene 91 for the double-bond position, upon interaction with (2*E*)-butenal 2 and benzaldehyde 8 in the presence of clay (Scheme 14) gives the same bicyclic products 5 and 27 as those formed in the reactions between these aldehydes and monoterpene 91 [23]. In both cases, the reaction presumably proceeds via the carbocation 6 (Scheme 1).

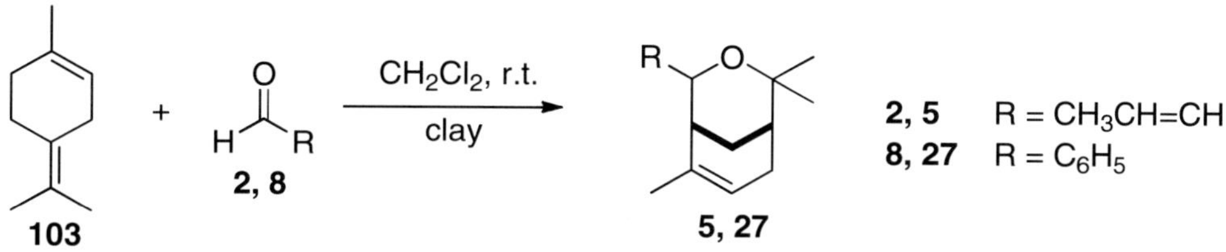

Scheme 14.

The authors of [28] studied the interaction between α-terpineol 104 and benzaldehyde 8 upon heating in the presence of $BF_3·Et_2O$, which gave heterocyclic compound 27 with a yield of 65% (Scheme 15). In accordance with the supposed mechanism [28], the reaction proceeds via the formation of hemiacetal 105. The same product 27, but with a lower yield of 45%, was obtained upon the interaction between α-terpineol 104 and benzaldehyde 8 in the presence of sulfuric acid [29].

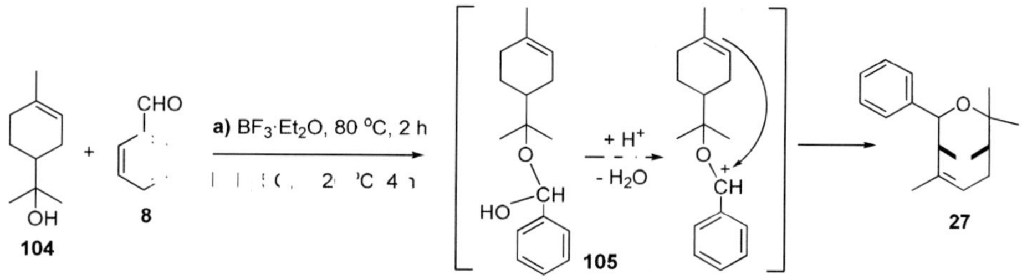

Scheme 15.

Trans-Sobrerol 106, as compared to α-terpineol 104, contains an additional hydroxyl group at position 8. Its interaction with aldehydes in the presence of $BF_3·Et_2O$ gives the products with the oxabicyclo[3.3.1]nonane framework with an additional keto group (Scheme 16) [30].

The yields of the products 116-119 and 122, which are obtained via the interaction between sobrerol 106 and aliphatic aldehydes 20-24 were considerably higher than those in case of using aromatic aldehydes, with the exception for benzaldehyde 8. It was of interest that the yield of heterocyclic compounds was virtually independent of the nature of substituents in the aromatic ring of the aldehydes.

A hypothesis was made [30] that carbocation 123 is formed at the first stage of the reaction; after the nucleophilic attack by an aldehyde this cation forms the oxocarbenium ion 124. This ion undergoes cyclization forming enol; the subsequent tautomerization in the resulting enol gives a heterocyclic product (Scheme 17).

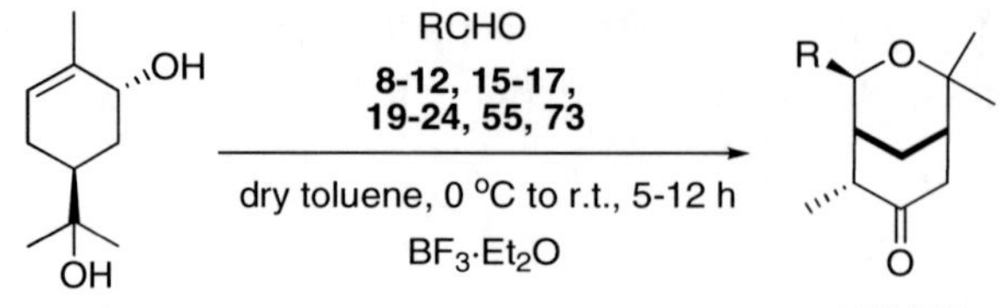

8, 107	R = C_6H_5 (80%)	**19, 115**	R = 4-Me-C_6H_4 (68%)
9, 108	R = 4-Cl-C_6H_4 (50%)	**21, 116**	R = C_2H_5 (98%)
10, 109	R = 4-Br-C_6H_4 (51%)	**22, 117**	R = C_3H_7 (98%)
11, 110	R = 4-F-C_6H_4 (62%)	**23, 118**	R = $(CH_3)_2CHCH_2$ (82%)
12, 111	R = 4-NO_2-C_6H_4 (50%)	**24, 119**	R = C_6H_{13} (80%)
15, 112	R = C_6H_5-CH=CH (57%)	**73, 120**	R = 4-MeO-C_6H_4 (56%)
16, 113	R = 4-NO_2-C_6H_4-CH=CH (62%)	**55, 121**	R = $C_6H_5CH_2$ (50%)
17, 114	R = 4-MeO_2C-C_6H_4 (52%)	**20, 122**	R = C_6H_{11} (90%)

Scheme 16.

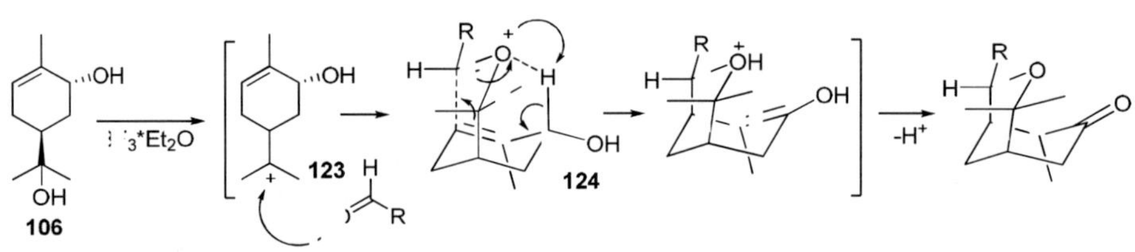

Scheme 17.

Reactions of 4-terpineol 125 with aldehydes, which are catalyzed by Lewis acids, yield products with the 6-oxabicyclo[3.2.1]octene framework (Scheme 18) [31].

Among the catalysts used (BF_3·Et_2O, $Sc(OTf)_3$, $In(OTf)_3$, $Bi(OTf)_3$, $InCl_3$), the best results have been achieved with $In(OTf)_3$, methylene chloride was the most suitable solvent.

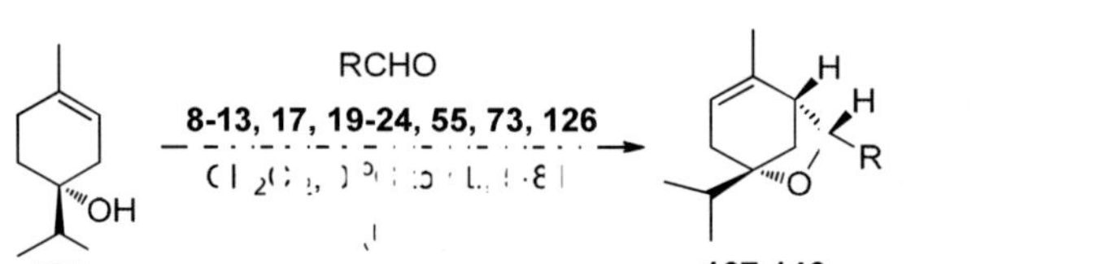

8, 127	R = C_6H_5 (75%)	**20, 135**	R = cyclo-C_6H_{11} (84%)
9, 128	R = 4-Cl-C_6H_4 (68%)	**21, 136**	R = C_2H_5 (94%)
10, 129	R = 4-Br-C_6H_4 (65%)	**22, 137**	R = C_3H_7 (92%)
11, 130	R = 4-F-C_6H_4 (66%)	**23, 138**	R = $(CH_3)_2CHCH_2$ (88%)
12, 131	R = 4-NO_2-C_6H_4 (62%)	**24, 139**	R = C_6H_{13} (85%)
13, 132	R = 2-NO_2-C_6H_4 (64%)	**73, 140**	R = 4-MeO-C_6H_4 (80%)
17, 133	R = 4-MeO_2C-C_6H_4 (60%)	**55, 141**	R = $C_6H_5CH_2$ (82%)
19, 134	R = 4-Me-C_6H_4 (82%)	**126, 142**	R = 3-Br-C_6H_4 (64%)

Scheme 18.

The yield of heterocyclic products in the reactions of aliphatic aldehydes with 125, as well as those in the reaction with sobrerol 106, turned out to be considerably higher than in the case of using aromatic aldehydes; however, the differences were less significant [31]. In general, the yields of these reactions with aromatic aldehydes containing the electron-donating substituents were higher compared to the aldehydes containing electron-withdrawing substituents.

The regularities being observed can presumably be attributed to the effect of stabilization of the oxocarbenium ion due to the donating properties of the aliphatic groups or substituents in the aromatic ring of aldehydes.

It was assumed [31] that the first stage of the reaction comprises the activation of an aldehyde with a Lewis acid for the nucleophilic attack by the hydroxyl group, giving acetal 143, which decomposes to yield the oxocarbenium ion 144.

The ion 144 subsequently undergoes cyclization, yielding products 127-142 (Scheme 19).

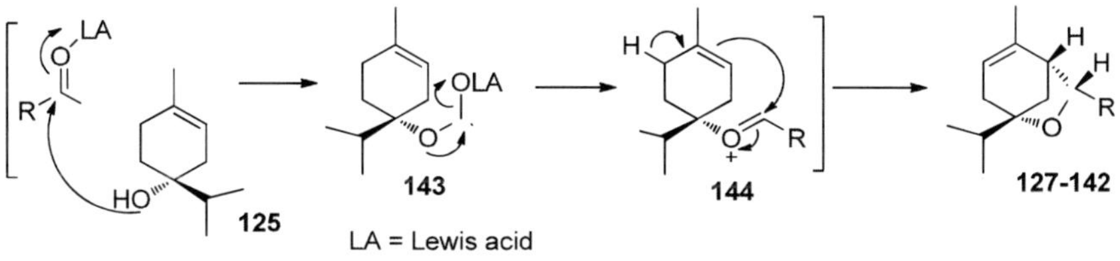

Scheme 19.

The interaction between isopulegol 71 and propanal 21 or isobutyral 145 without a solvent in the presence of catalytic amounts of *para*-toluenesulfonic acid applied onto silica gel results in the formation of diastereomeric products with the chromene framework 81 and 146 (Scheme 20) [32].

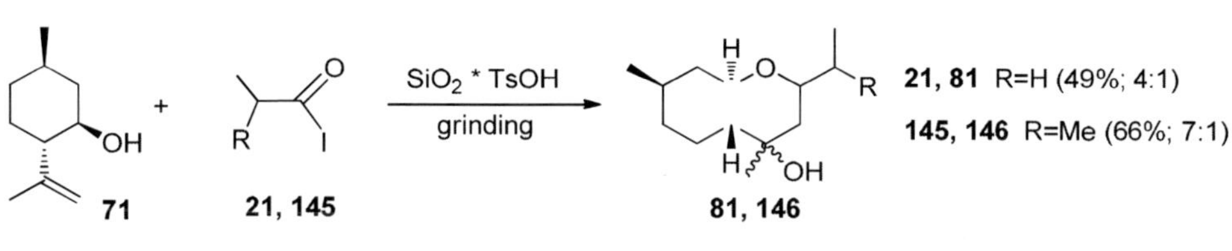

Scheme 20.

The interaction between isopulegol 71 and 4-methoxybenzaldehyde 73 in the presence of I_2 also yields the mixture of chromene products 82a,b at a 5:1 ratio (Scheme 21). The authors of [33] believe that the key stage of this reaction is the attachment of a hydroxyl group of isopulegol 71 to the aldehyde activated by the interaction with I_2 (147), which leads to the formation of cation 148. The subsequent transformations may proceed in accordance with the mechanism shown in Scheme 10.

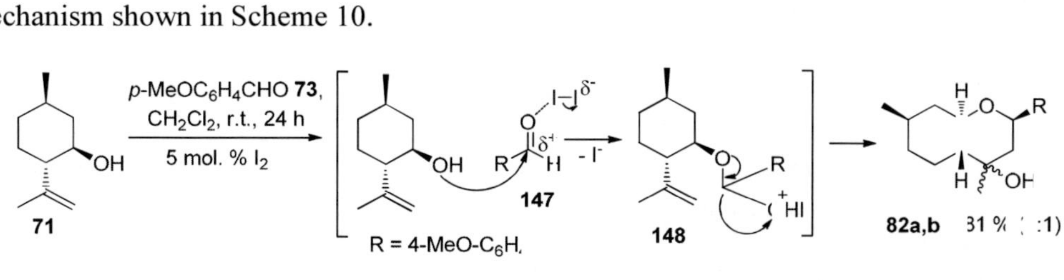

Scheme 21.

The interaction between isopulegol 71 and aldehydes 10 and 22 carried out in the presence of $BF_3{\cdot}Et_2O$ in benzene resulted in the formation of compounds 149 and 150 with the yields of 60 and 64%, respectively (Scheme 22) [34].

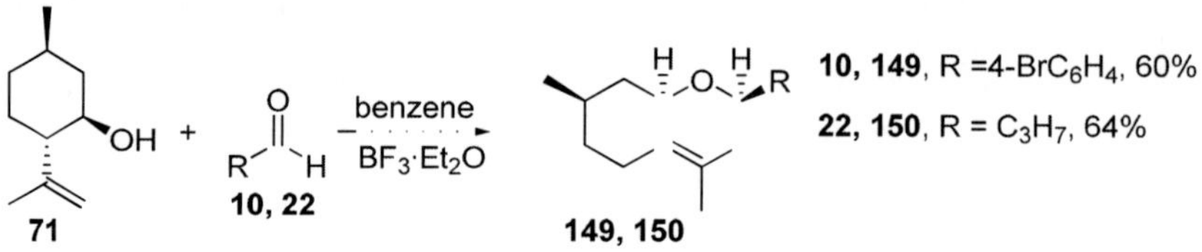

Scheme 22.

6-Hydroxymethyl-limonene 151 interacts with salicylaldehyde 47 in the presence of montmorillonite clay to form a tricyclic product 152 with the yield of 34% (Scheme 23) [6, 24]. The key stage of the tentative mechanism of formation of compound 152 is carbocyclization into cation 153. It is an interesting fact that unlike the reactions of aldehyde 47 with geraniol 7 (Scheme 4) and limonene 91 (Scheme 13), the phenolic hydroxyl group remains intact in this case.

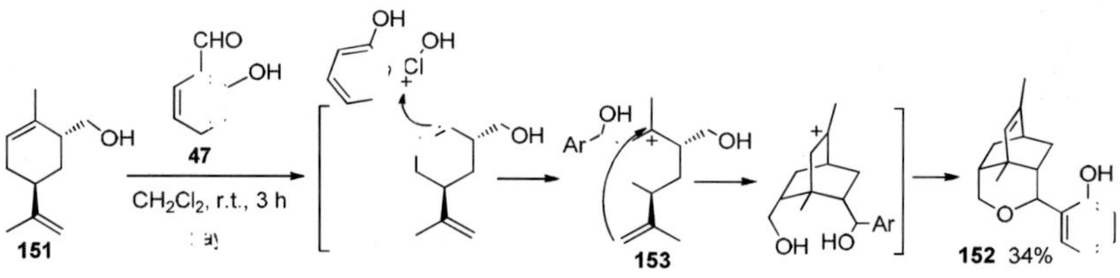

Scheme 23.

Another alcohol with the *para*-menthane framework, 6-hydroxymethyl-*iso*-limonene 154, which is the isomer of 6-hydroxymethyl-limonene 151, yields product 155 with a completely different type of framework in the reaction with salicylaldehyde 47 (Scheme 24) [6, 24].

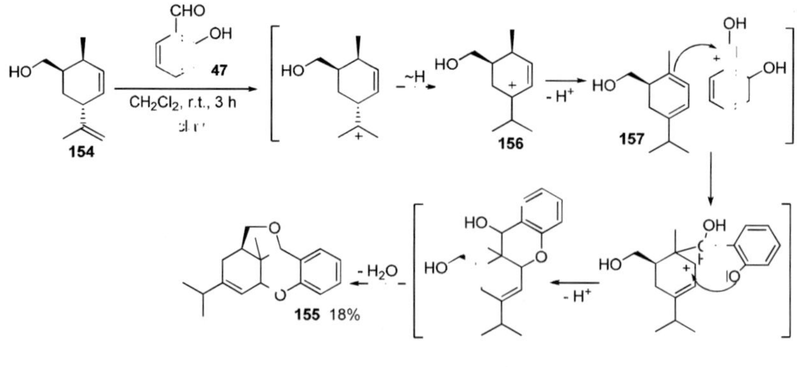

Scheme 24.

The authors [24] assume that in the first step, protonation of exocyclic double bond occurs, followed by a 1,2-hydride shift to give the ion 156. After loss of a proton the intermediate 157 reacts with the protonated aldehyde.

The interaction between terpinolene epoxide 158 and aliphatic aldehydes 21, 145, and 159 in the presence of $BF_3 \cdot Et_2O$ was studied in [35]. Acetals 160-162 were the main products of the reaction (Scheme 25).

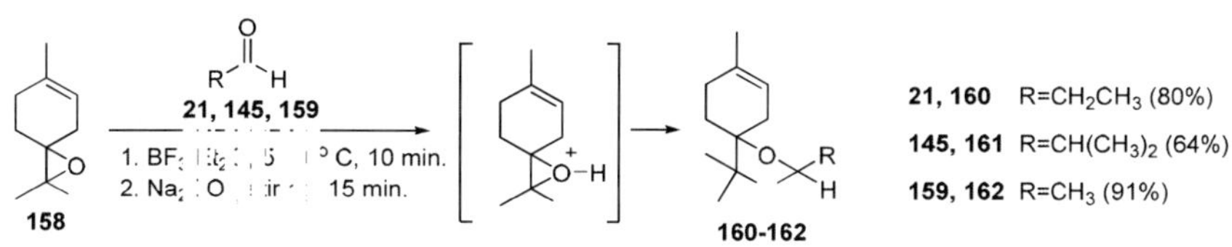

Scheme 25.

After the interaction between the mixture of *cis*- and *trans*-limonene-1,2-epoxides 163 and 164 (at the 1:1 ratio) and methacrylaldehyde 93 on askanite-bentonite montmorillonite clay at 0°C, the unreacted cis-epoxide 163, acetal 165 and diol 166 were isolated from the reaction mixture (Scheme 26) [36]. The interaction between the individual epoxide 163 and aldehyde 93 gave diol 166, ketone 167, and acetal 168 as the main products [36].

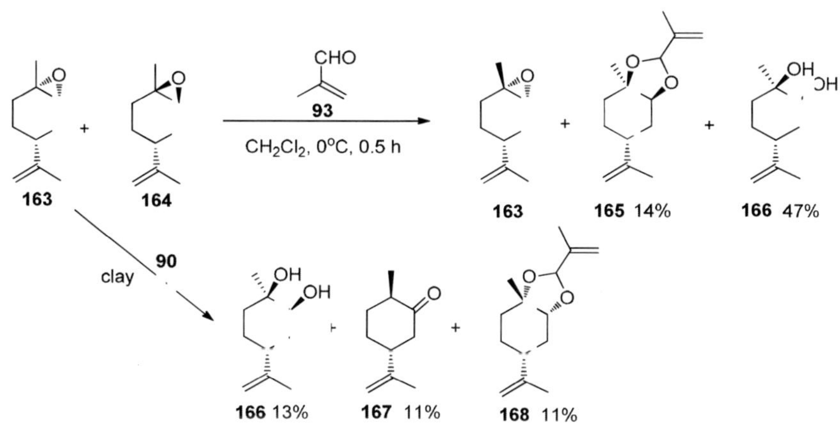

Scheme 26.

Diol 169 with the *para*-menthane framework in the presence of montmorillonite clay K10 interacts with crotonic aldehyde 2 and aromatic aldehydes 170 and 171 containing several electron-donating substituents, which gives a number of heterocyclic compounds 172-178 with different types of frameworks (Scheme 27) [37]. The main products of all the reactions were compounds with the chromene framework; the distribution of side products depended on the features of aldehyde structure. Let us note that products 172a,b exhibit high analgesic activity *in vivo* [38].

The authors of [37] assumed that the products with the chromene framework (172a,b, 174a,b, and 177a,b) were formed due to the interaction between the protonated aldehyde and a double bond of terpenoid 169, which leaded to the formation of an intermediate carbocation 179. The intramolecular heterocyclization in this cation gives diastereomeric products (Scheme 28).

The supposed mechanism of formation of products with the benzodioxin framework (173 and 176) includes protonation of diol 169 and formation of carbocation, which interacts with the aldehyde, which plays a role of a nucleophile.

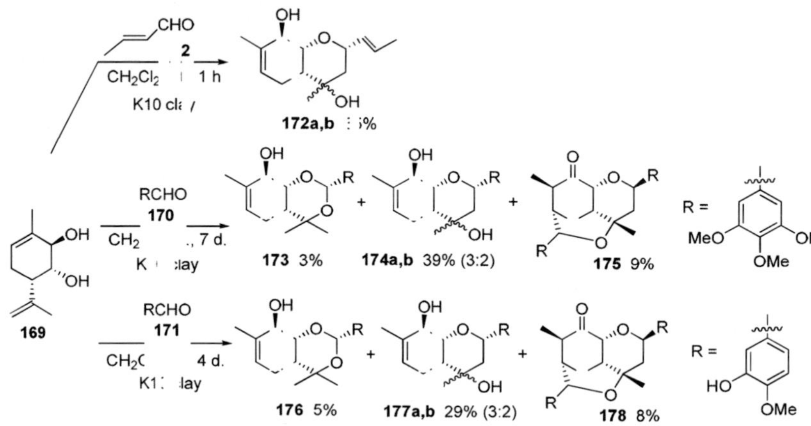

Scheme 27.

Heterocyclization can subsequently take place in the resulting cation 180, giving compounds 173 and 176 (Scheme 28).

The formation of tricyclic products 175 and 178 may comprise binding of the protonated aldehyde to the double bond of compound 169 and the subsequent heterocyclization [37]. Then, the resulting cation 181 interacts with the next aldehyde molecule. The subsequent proton detachment and tautomerization give products, such as 175 and 178 (Scheme 28).

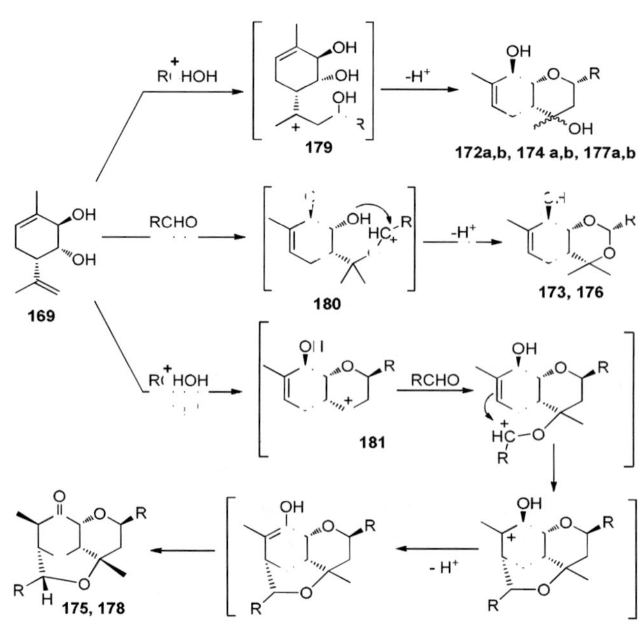

Scheme 28.

Interaction of Aldehydes with Bicyclic Monoterpenoids

Interaction of Aldehydes with Monoterpenoids with Pinane Framework

α- And β-pinenes 182 and 183 are the major components of coniferous resin. α-Pinene 182 is available in form of both enantiomers; whereas β-pinene 183 is present in significant amounts in natural sources only in the form of the (-)-enantiomer.

Under acidic conditions, pinenes and their derivatives can be easily rearranged to form compounds with the *para*-menthane framework, which is the reason why a number of products of the reactions with aldehydes are identical for these two types of monoterpenoids.

Under conditions of acidic heterogeneous catalysis, β-pinene 183 interacts with aliphatic aldehydes (2*E*)-butenal 2 and butyraldehyde 22, which gives the compounds 5 and 41 with 3-oxabicyclo[3.3.1]non-6-en framework (Scheme 29) [39]. As mentioned previously, the same products can be obtained via the interaction between aldehydes and myrcene 1 (Scheme 1), geraniol 7 (Scheme 2), or limonene 91 (Scheme 12). The reaction in all these cases presumably proceeds via the same intermediate carbocation 6.

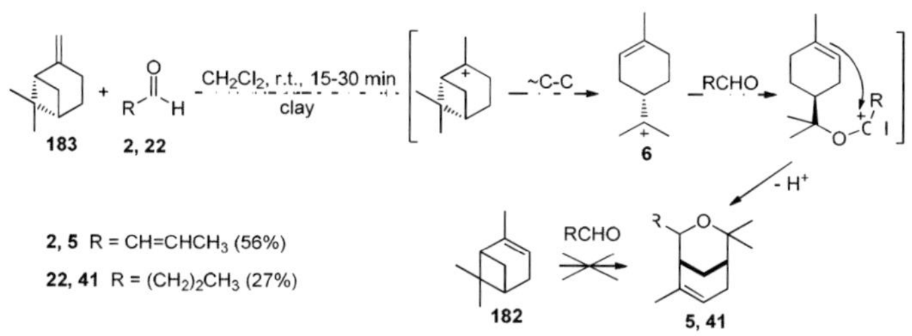

Scheme 29.

It is an interesting fact that α-pinene 182, which is supposed to form the same tertiary carbocation 6 as that formed by β-pinene 183, does not interact with the aforementioned aldehydes in the presence of clay [39].

Meanwhile, (-)-α-pinene epoxide 184 in the presence of montmorillonite clay interacts with (2*E*)-butenal 2 and methacrylaldehyde 93, giving the products 185 and 186, which also have the oxabicyclo[3.3.1]nonane framework with an additional keto group (Scheme 30) [40]. The compounds of the same structural type can be obtained via the interaction between sobrerol 106 and aldehydes (Scheme 16).

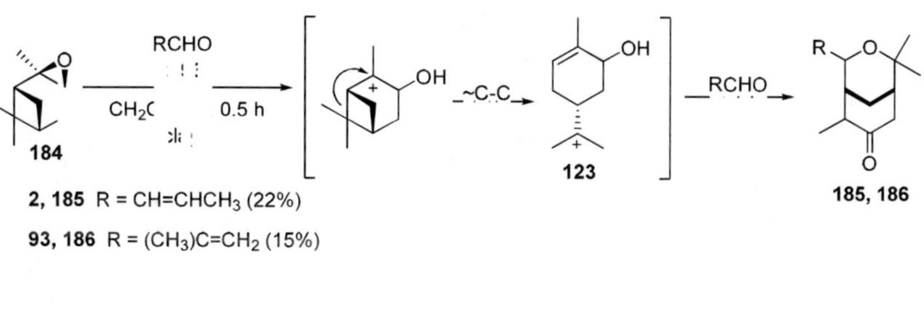

Scheme 30.

The opening of the protonated epoxide cycle is presumably followed by the rearrangement into cation 123 with the *para*-menthane framework [40], which can subsequently interact with aldehydes, as it is shown in Scheme 17 by an example of the reaction between aldehydes and sobrerol 106.

The interaction between epoxide 184 and salicylaldehyde 47 is accompanied by deep rearrangement of the pinane framework, giving compound 187, which contains the coupled seven-, six-, five-, and four-membered rings [40] (Scheme 31). Cation 188 with the cyclopentane framework and aldehyde 189 are tentatively the key intermediates in this extremely unusual reaction. Despite the fact that the all asymmetric centers, which are available in the initial molecule 184, participate in the reaction, the transformations occur stereoselectively yielding only one stereoisomer [40].

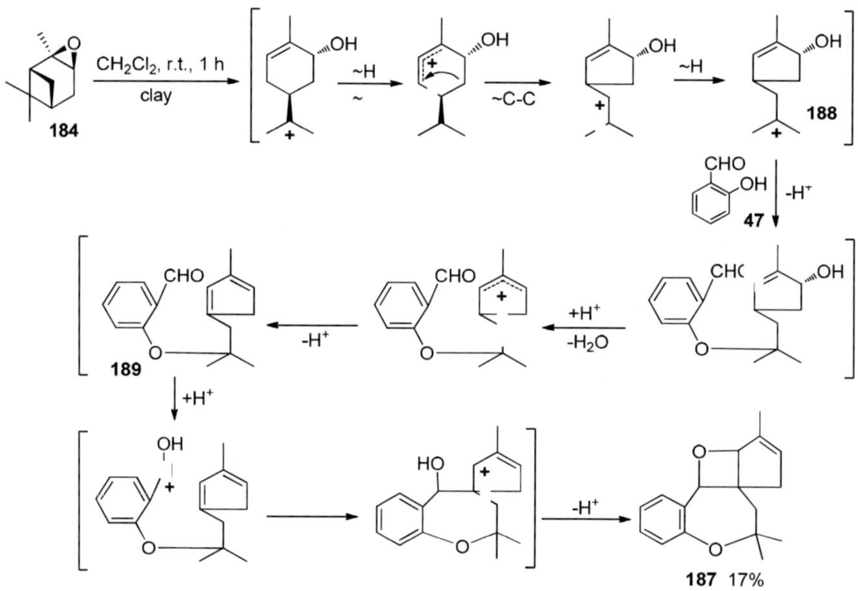

Scheme 31.

The interaction between β-pinene epoxide 190 and acrylaldehyde 92 in the presence of clay gives the product 191 with the 3-oxabicyclo[3.3.1]nonane framework, which contains the aldehyde group, and two stereoisomers of tricyclic spirocompounds 192a,b (Scheme 32) [40]. The supposed mechanism of these transformations comprises the opening of the protonated epoxide group, rearrangement of the cation 193 with the pinane framework into the corresponding cation with the *para*-menthane framework, and the interaction of the latter with a molecule of aldehyde 92, giving intermediate 194.

Cation 194 subsequently can either be transformed into aldehyde 191 or interact with another acrolein molecule, yielding the mixture of acetals 192a,b.

During its interaction with salicylaldehyde 47 in the presence of clay, (+)-*trans*-verbenol (+)-195 participates in the double heterocyclization reaction, which is accompanied by skeletal rearrangement and eliminating of a water molecule, yielding tetracyclic product 196 (Scheme 33) [41]. Reactions of the compound (+)-195 with (2*E*)-butenal 2 and methacrylaldehyde 93 on clay give compounds (+)-197 and (+)-198. The structures of the resulting products are similar to those of compounds 5 and 98 synthesized via the interaction of these aldehydes with limonene 91 (Schemes 12, 13); however, they contain an additional double bond.

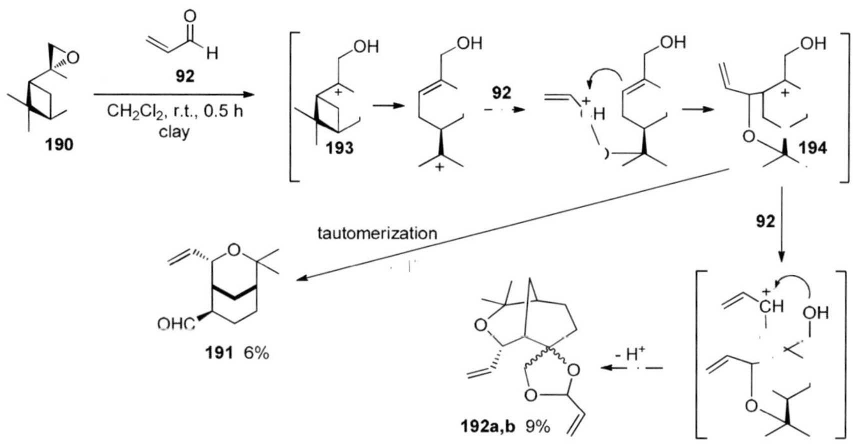

Scheme 32.

The interaction of (-)-*cis*-verbenol (-)-195 with aldehydes 2 and 47 on askanite-bentonite clay resulted in the formation of compounds (-)-197 and (-)-196, respectively, which are the optical antipodes of heterocyclic compounds synthesized from (+)-*trans*-verbenol (+)-195 [41]. Thus, the relative position (*cis*- or *trans*-) of the hydroxyl group in verbenol 195 does not affect the direction of the transformations taking place.

In the presence of montmorillonite clay, (-)-*cis*-verbenol epoxide 199 interacts with the aliphatic ((2*E*)-butenal 2) and aromatic (8-12, 18, 47, 73, 170, 171, 202) aldehydes [37, 41, 42]. The competition between the intramolecular transformations giving compounds 169, 200 and 201, and intermolecular reactions is an important feature of these reactions (Scheme 34, Table 2).

The ratio between the intra- and intermolecular processes and the distribution of heterocyclic products is determined by the structure of the initial aldehyde. Thus, when (2*E*)-butenal 2 is used as the initial compound, compound 203 with the benzodioxine type of framework turned out to be the only intermolecular product isolated [41].

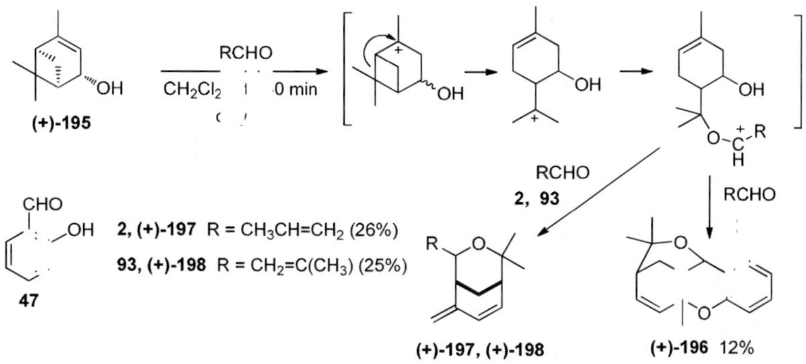

Scheme 33.

The reaction between benzaldehyde 8 and monoterpenoid 199 gives products 204 and 212a,b with the benzodioxine and chromene frameworks at a comparable ratio [37]. Meanwhile, the interaction between epoxide 199 and almost all *para*-substituted benzaldehydes (9-12 and 73) yields exclusively the compounds with the benzodioxine framework as intermolecular products. It should be noted that when the aldehydes containing the electron-withdrawing group (-Cl, -Br, -I, $-NO_2$) were used, a decrease in the yield of intermolecular products was observed [42].

Compounds 205, 207, and 208 with the benzodioxine framework demonstrated high analgesic activity *in vivo* [43].

Although the reaction of (–)-cis-verbenol epoxide 199 with 4-methoxybenzaldehyde 73 gave 13% of compound 210, only traces of the corresponding product were detected in the reaction mixture obtained with *para*-dimethylaminobenzaldehyde 216.

Thus, the reaction of 199 with *para*-substituted benzaldehydes is governed not only by donor–acceptor properties of the *para*-substituent but also by specificity of adsorption of the initial aldehyde on K10 [42].

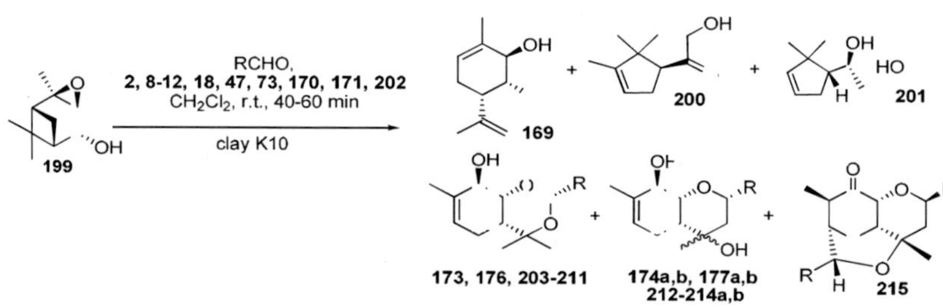

2, 203	R = $CH_3CH{=}CH$	18, 213a,b	R = 4-OH-C_6H_4
8, 204, 212a,b	R = C_6H_5	47, 209	R = 2-OH-C_6H_4
9, 205	R = 4-Cl-C_6H_4	73, 210	R = 4-OMe-C_6H_4
10, 206	R = 4-Br-C_6H_4	170, 173, 174a,b, 215	R = 3,4,5-OMe-C_6H_2
11, 207	R = 4-F-C_6H_4	171, 176, 177a,b	R = 3-OH-4-OMe-C_6H_3
12, 208	R = 4-NO_2-C_6H_4	202, 211, 214a,b	R = 4-OH-3-OMe-C_6H_3

Scheme 34.

Table 2. Reactions of (–)-*cis*-verbenol epoxide 199 with aldehydes

No	Aldehyde (R)	Products yield, % (No)					
		Intramolecular			Intermolecular		
		169	200	201	Dioxin type	Chromene type	215
2	$CH_3CH{=}CH$-	13	5	-	16 (203)	-	-
8	C_6H_5-	16	7	-	8 (204)	6 (212a,b)	-
9	4-Cl-C_6H_4-	27	10	4	8 (205)	-	-
10	4-Br-C_6H_4-	45	14	4	6 (206)	-	-
11	4-F-C_6H_4-	26	11	2	15 (207)	-	-
12	4-NO_2-C_6H_4-	34	13	8	5 (208)	-	-
18	4-OH-C_6H_4-	34	14	-	-	8 (213a,b)	-
47	2-OH-C_6H_4-	51	-	-	4 (209)	-	-
73	4-OMe-C_6H_4-	33	7	-	13 (210)	-	-
170	3,4,5-OMe-C_6H_2-	10	12	-	6 (173)	25 (174a,b)	5
171	4-OMe-3-OH-C_6H_3-	-	15	5	5 (176)	28 (177a,b)	-
202	4-OH-3-OMe-C_6H_3-	28	14	-	10 (211)	25 (214a,b)	-

It is an interesting fact that compounds 213a,b with the chromene framework [37] were the only intermolecular product isolated upon the interaction between epoxide 199 and 4-hydroxybenzaldehyde 18; but compound 209 with the benzodioxine framework was the only heterocyclic compound isolated in the reaction with salicylaldehyde 47 [41].

The interaction of verbenol epoxide 199 with isovanillin 171 and vanillin 202 in the presence of clay resulted in the preferential formation of products 177a,b and 214a,b, respectively, which had the chromene framework. The yield of compounds 176 and 211 with the benzodioxine framework was lower [37]. Note that products 214a,b with the chromene framework exhibited high analgesic activity in the *in vivo* experiments [44].

In the reaction of 3,4,5-trimethoxybenzaldehyde 170 with epoxide 199, compound 215, which is presumably formed via the attachment of two aldehyde 170 molecules to terpenoid 199, was isolated along with products 173 and 174a,b [37].

The mechanisms of formation of intermolecular heterocyclic products in the reactions of (–)-*cis*-verbenol epoxide 199 with aldehydes are presumably similar to the same transformations for diol 169, which are shown on Scheme 28, with the exception for the preliminary stage of the rearrangement of epoxide 199 into cation with the *para*-menthane framework 217 or compound 169 (Scheme 35). It is noteworthy that the reactions between aromatic aldehydes and epoxide 199 proceed much faster (reaction duration is less than 1 h) as compared to the same transformations with compound 169, where it takes several days to attain complete conversion (Scheme 27).

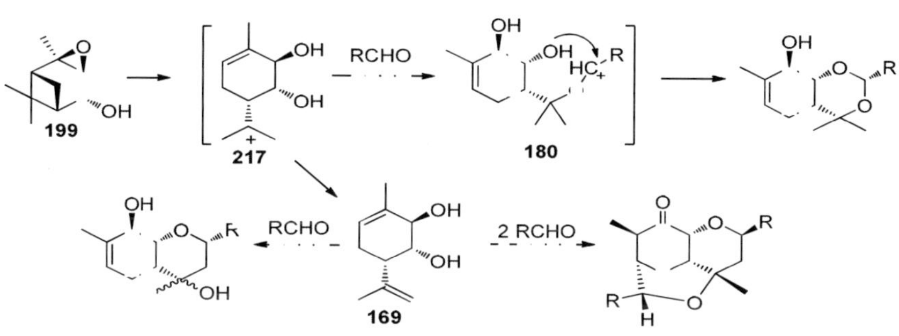

Scheme 35.

The reactions between verbenone epoxide 218 and aliphatic aldehydes (2*E*)-butenal 2 и acrylaldehyde 92 in the presence of clay yield diketones with the 3-oxabicyclo[3.3.1]non-6-enon framework 219 and 220, which exist partly in the enol form (Scheme 36) [45]. The reaction presumably proceeds via the carbocation 221 obtained as a result of opening of the cyclobutane ring, followed by the interaction with aldehydes.

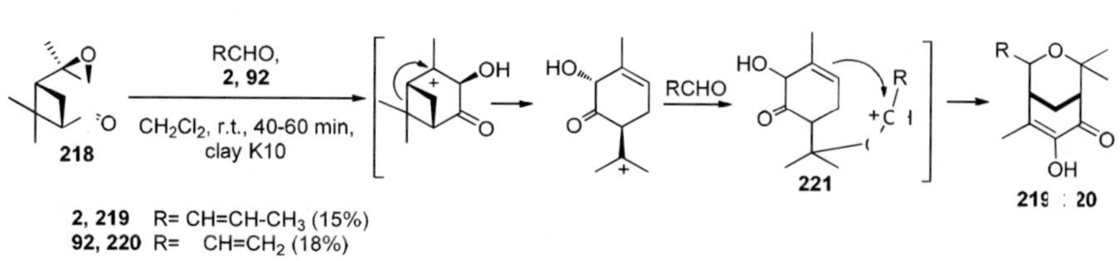

Scheme 36.

The interaction of verbenone epoxide 218 and salicylaldehyde 47 on clay results in the formation of the mixture consisting of the isomerization product ketoalcohol 223 and heterocyclic compound 222 (Scheme 37) [45]. It is curious that the interaction of terpenoid 218 with salicylaldehyde 47 does not proceed via the double heterocyclization yielding tetracyclic products as it was for the reactions of aldehyde 47 with geraniol 7 (Scheme 4) and limonene 91 (Scheme 13).

The reaction between verbenone epoxide 218 and 4-methoxybenzaldehyde 73 in the presence of clay results in the formation of a mixture of tautomeric bicyclic products 224 and 225 with the 3-oxabicyclo[3.3.1]non-6-en framework, as well as compound 226, which is presumably formed as a result of a ring contraction in cation 227 [45].

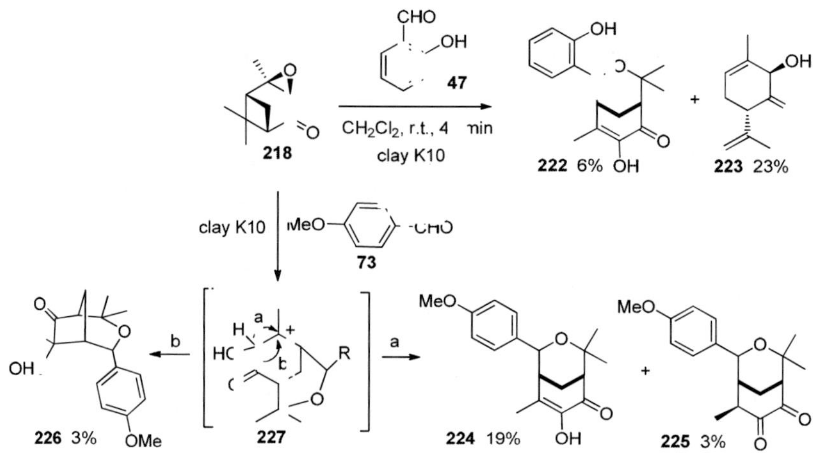

Scheme 37.

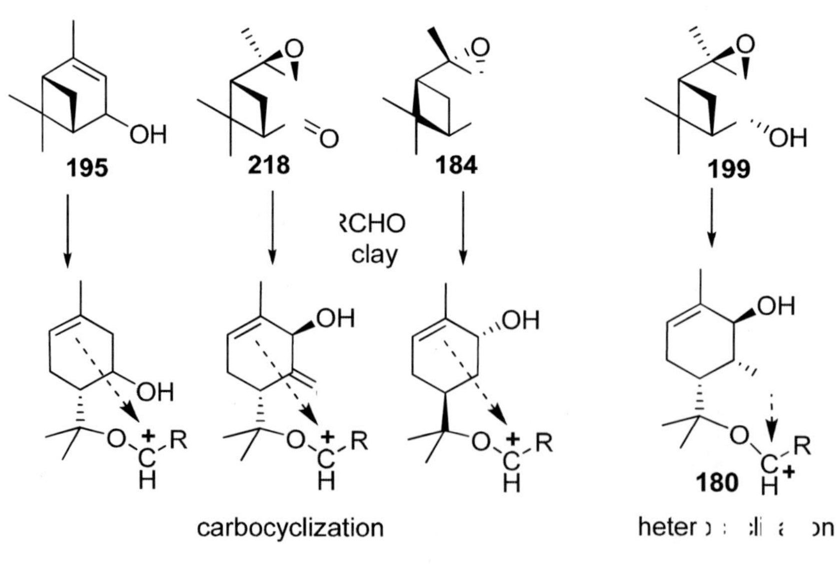

Scheme 38.

It is clear from Scheme 38 that the structures of the cations, which can be formed from verbenone epoxide 218, *trans*- and *cis*-verbenols 195, α-pinene epoxide 184 upon their interaction with aldehydes on clay are very similar to that of cation 180, which is presumably formed from *cis*-verbenol epoxide 199. Nevertheless, whereas the products that can result exclusively from the occurrence of carbocyclization at the final stage of the reaction were obtained in the four cases, verbenone epoxide 199 yielded only the heterocyclization product. Therefore, even relatively small changes in terpenoid structure in the case of using montmorillonite clay as a catalyst may result in an abrupt change in the direction of the reaction [41].

INTERACTION OF ALDEHYDES WITH MONOTERPENOIDS WITH CAMPHANE FRAMEWORK

Camphene 228 is the key intermediate product in the synthesis of camphor. Although camphene 228 rather frequently occurs in essential oils, its content is typically small. The major methods for the synthesis of this compound are rearrangements of α- and β-pinenes 182 and 183 under acidic conditions.

In the presence of acidic catalysts ($SnCl_4$, H_3PO_4, AcOH, $BF_3{\cdot}Et_2O$), camphene 228 participates in a condensation reaction with formaldehyde 229 (Prins reaction), yielding 8-hydroxymethyl-camphene 230 or its ethers (Scheme 39) [46]. When carried out in the presence of cation exchange resin KU-2, the same reaction gives alcohol 230 and tricyclic product 231 (Scheme 39) [47].

When performing the reaction between formaldehyde 229 and camphene 228 on β-zeolite, the formation of the products of skeletal rearrangements, tricyclic compounds 232 and 233, is the main direction (Scheme 39) [48]. The first stage of the supposed mechanism comprises the attack of the protonated aldehyde 229 at olefin double bond of compound 228; the reaction subsequently proceeds along two pathways: a) 1,2-*exo*-methyl shift and cyclization giving compound 232; b) Wagner–Meerwein rearrangement and intramolecular cyclization yielding product 233.

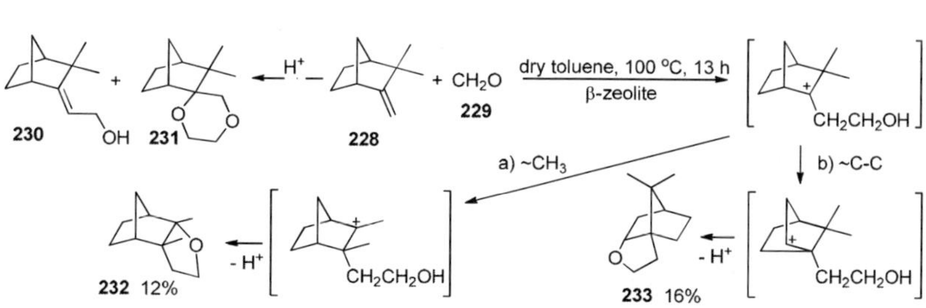

Scheme 39.

Another bicyclic monoterpenoid with the bicyclo[2.2.1]heptane framework, α-fenchene 234, which differs from camphene 228 only by the position of methyl groups, in the presence of β-zeolite interacts with benzaldehyde 8, giving a mixture of diastereomers 235a,b and isomeric product 236 (Scheme 40) [49]. The reaction presumably is initiated by the protonated aldehyde attacking the double bond of α-fenchene 234, followed by the Wagner–

Meerwein rearrangement in the resulting cation, which can be accompanied by a hydride shift followed by heterocyclization [49].

Camphene epoxides 237a,b interact with unsaturated aldehydes (acrylaldehyde 92, methacrylaldehyde 93 and (2*E*)-butenal 2) in the presence of montmorillonite clay to form acetals 238-240 with the yields of 11–18% and diastereomeric aldehydes 241a,b (Scheme 41) [50]. The presumable mechanism includes opening of the protonated epoxide cycle to form the cation 242, which can either be rearranged into compounds 241a,b or be bound with the aldehyde molecule, giving acetals 238-240.

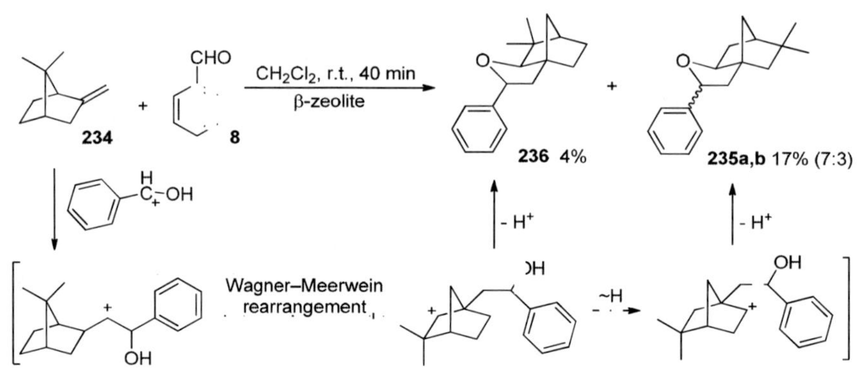

Scheme 40.

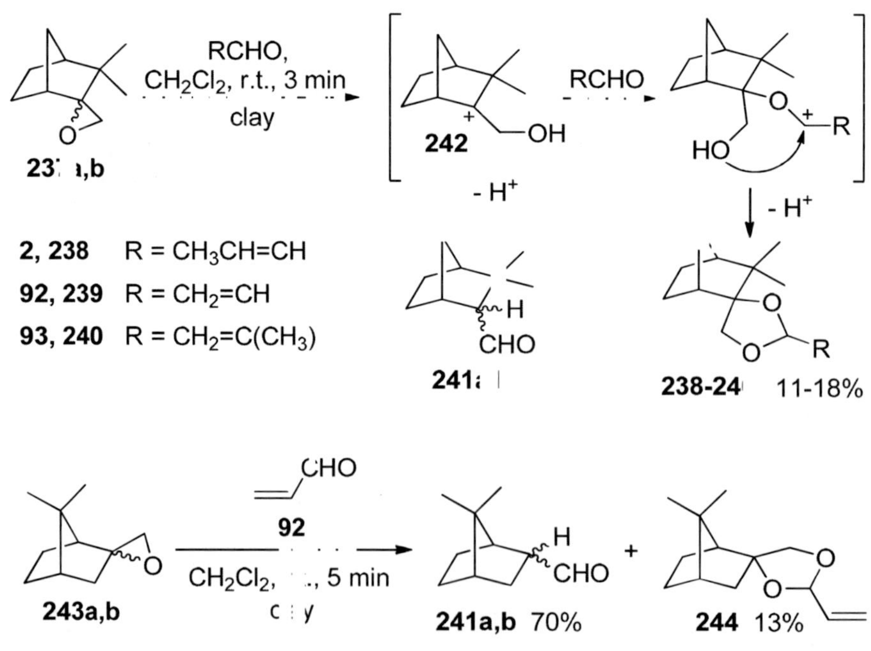

Scheme 41.

The same transformations occur upon the interaction between α-fenchene epoxides 243a,b and acrolein 92 in the presence of clay, resulting in the formation of compound 244 as a heterocyclic product.

INTERACTION OF ALDEHYDES WITH MONOTERPENOIDS CONTAINING CYCLOPROPANE RING

(+)-3-Carene 245 occurs in pine turpentine in appreciably large amount and is characterized by high optical purity. Although its isomer (+)-2-carene 246 is also abundant in nature, its content is essential oils is extremely small, and the isolation of (+)-2-carene 246 from the oils is a difficult task. (+)-2-Carene 246 is typically obtained via the isomerization of (+)-3-carene 245 in the presence of strong bases at a high temperature: an equilibrium mixture of 2- and 3-carenes at the 2:3 ratio is formed under these conditions [51, 52]. The subsequent isolation of the individual 2-carene 246 is carried out by distillation of the mixture on a rectification column.

The reaction of (+)-2-carene 246 with (2*E*)-butenal 2 in the presence of K10 clay for 3 days led to products of intermolecular interaction, namely, compounds 247a,b with the isobenzofuran framework at a ratio of 1:1 (Scheme 42) [53].

The hypothetical mechanism responsible for the formation of 247a,b includes an attack of the protonated aldehyde at the cyclopropane ring of 246 followed by intramolecular cyclization.

Note that the reaction of terpene 246 with (2*E*)-butenal 2 forms only the products corresponding to the attack of the protonated aldehyde at the cyclopropane ring but not at the C=C bond.

The *cis*-condensation of the rings in compounds 247a,b was an unexpected result. One of the possible explanations to this fact is an assumption that the formation of the intermolecular product upon interaction between 2-carene 246 and aldehyde 2 requires their adsorption at closely located catalytic centers (e.g., in the interlayer space of the clay), which disables the attack resulting in the formation of the product with *trans* condensation of the rings (Scheme 42) [53].

The reactions of 246 with 4-methoxybenzaldehyde 73 and vanillin 202 on K10 clay gave compounds 248a,b and 249a,b, which were analogs of 247a,b, and a smaller amount of compounds 96 and 250 with the 3-oxabicyclo[3.3.1]nonane framework (Scheme 43). On the reaction pathway to 96 and 250, carbocation 6, which is formed by the protonation of 246 at the cyclopropane ring accompanied by isomerization to the *para*-menthane framework, is probably the electrophile, and the aldehyde is the nucleophile, in contrast to the mechanisms described on Scheme 42.

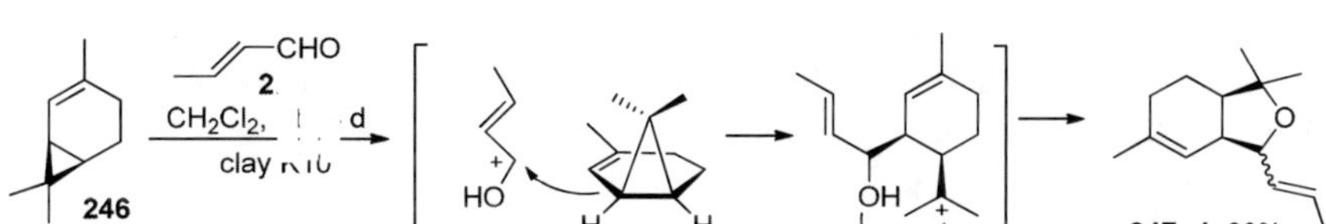

Scheme 42.

The reactions of 3-carene 245 with aldehydes 73 and 202 in the presence of K10 clay gave the same products as from 246, but in much lower yields (2–3% each). Compounds 248a,b cannot be formed from 245 without C=C bond migration. Since C=C bond migration in the product is improbable, it seems that 245 is initially partly rearranged on clay into 246, which then reacts with the aldehydes [53].

Carenes 245 and 246 did not react with benzaldehyde 8, 2-hydroxybenzaldehyde 47, 4-bromobenzaldehyde 10 and 4-fluorobenzaldehyde 11 in the presence of clay.

Since the reactions of carenes 245 and 246 with 4-methoxybenzaldehyde 73 on K10 clay form the same products, the equilibrium mixture of 245 and 246 at a ratio of 2:3 was used in the reaction with aldehyde 73 in the presence of K10 clay, giving 248a,b and 96 with yields of 35% and 2%, respectively [53]. Thus, using a mixture of 245 and 246 allowed obtaining unexpectedly high yields of compounds 248a,b because if we take into account the content of (+)-2-carene 245 in the starting mixture, the yield of 248a,b is 88%. One of the possible explanations is provided by the assumption that the side reactions of isomerization and oligomerization of (+)-2-carene 246 are largely suppressed in the presence of (+)-3-carene 245. The interaction of 3-carene *trans*-epoxide 251 with methacrylaldehyde 93 in the presence of montmorillonite clay gives the isomerization product (aldehyde 252 with the cyclopropane ring), acetal 253, and diol 254 [36] (Scheme 44). The presumable mechanism of these transformations includes the interaction between carbocation 255 that is formed via the opening of the protonated epoxide and the aldehyde group of methacrylaldehyde 93, giving intermediate 256 and subsequently acetal 253. Hydrolysis of acetal 253 yields diol 254.

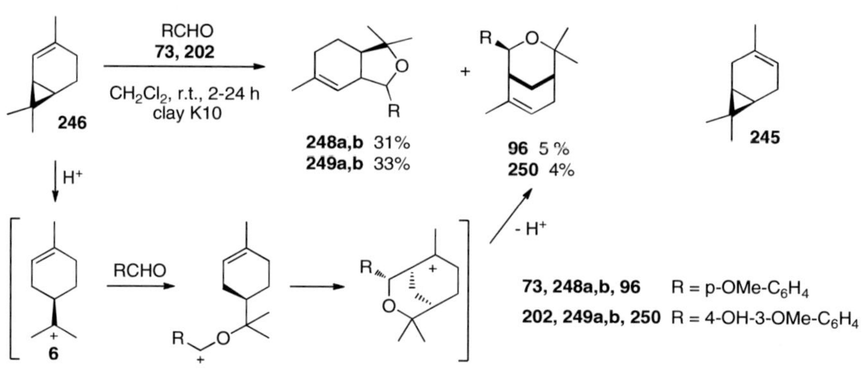

Scheme 43.

The reaction of 3-carene *cis*-epoxide 257 with methacrylaldehyde 93 gives another set of products, including aldehyde 258, ketone 259, and acetal 260 with the cleaved cyclopropane ring (Scheme 45) [36]. Thus, the relative spatial arrangement of the cyclopropane and epoxide rings has the key effect on the transformation pathways of epoxides 251 and 257 in the presence of clay.

The main product of the Prins reaction between 3-carene 245 and formaldehyde 229 is 4-hydroxymethyl-2-carene 261 [54], which, as opposed to 3-carene 245, exhibited a rather high reactivity in the interactions with aromatic and aliphatic aldehydes.

The reaction of *trans*-4-hydroxymethyl-2-carene 261 with salicylaldehyde 47 in the presence of montmorillonite clay gives product 155 with the xanthene framework condensed with the tetrahydrofuran ring with the yield of 62% (Scheme 46) [6]. The same compound (although with a considerably lower yield) is formed via the interaction between 6-hydroxymethyl-*iso*-limonene 154 and aldehyde 47 (Scheme 24).

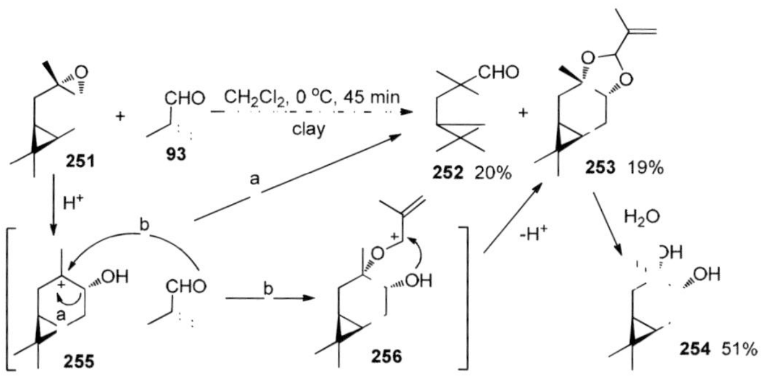

Scheme 44.

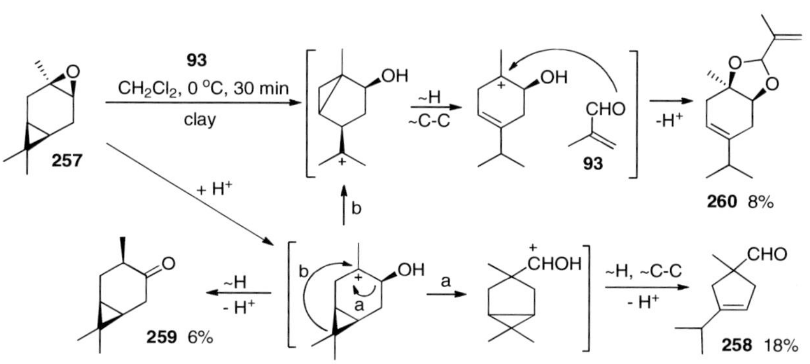

Scheme 45.

The presumable mechanism of this interaction includes the attack of the protonated aldehyde at the double bond of hydroxyolefin 261, the formation of a relatively stable α-cyclopropylcarbinyl ion 262, and subsequent heterocyclizations yielding product 155 [6].

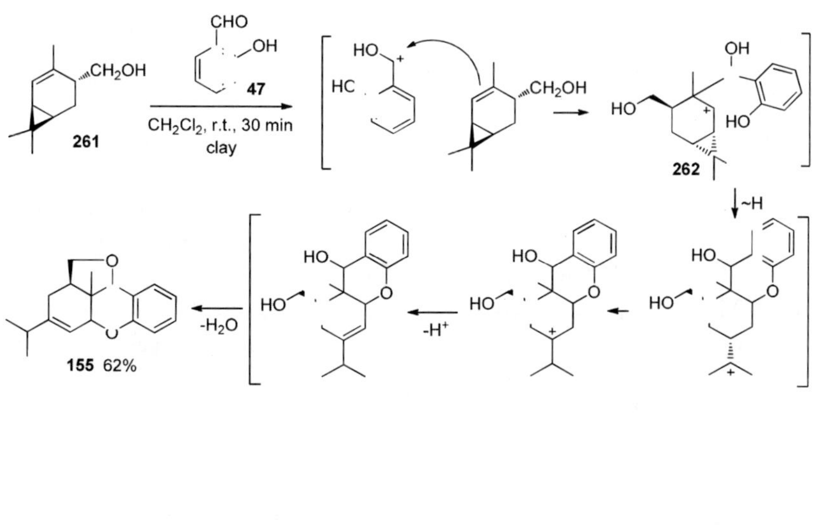

Scheme 46.

The attack of the protonated aldehyde at the double bond of *trans*-4-hydroxymethyl-2-carene 261 presumably initiates the reaction between compound 261 and benzaldehyde 8 in the presence of clay, resulting in the formation of isomeric isobenzofurans 263a,b and alcohol 264 with the *para*-menthane framework (Scheme 47) [6].

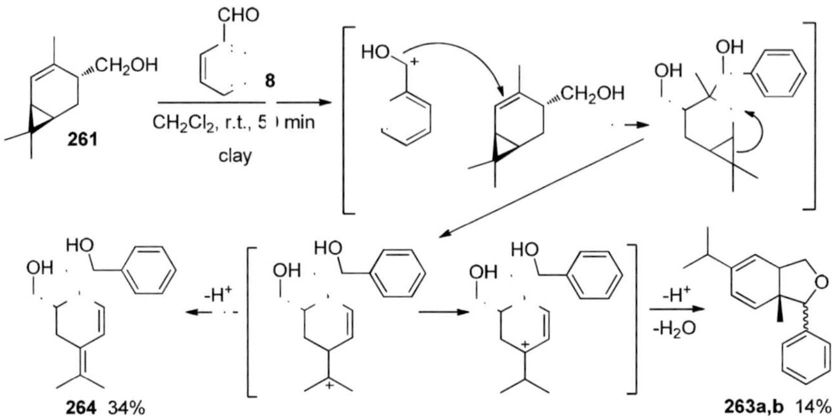

Scheme 47.

The reaction of *trans*-4-hydroxymethyl-2-carene 261 with aliphatic aldehydes (2*E*)-butenal 2, butanal 22, acrylaldehyde 92, methacrylaldehyde 93, and cinnamaldehyde 15, which contains an aromatic ring, in the presence of montmorillonite clay gives a broad range of intermolecular products containing at least two cycles [55].

The reaction of 261 with butanal 22 yielded compounds 265 and 266a,b with the isobenzofuran framework and compound 267, which was the product of binding of two aldehyde 22 molecules to monoterpenoid 261, as the main product (Scheme 48).

The supposed mechanism of formation of compounds 265 and 266a,b coincides with that for compounds 263a,b and includes the electrophilic attack of the protonated aldehyde at the double bond of hydroxyolefin 261 [55]. The transformations yielding compound 267 proceed via a different pathway. The reaction is presumably initiated by the attack of the protonated aldehyde at the cyclopropane ring of compound 261 rather than at the double bond, as it occurred for the previously described transformations. Cation 268, which is formed as a result of this interaction, undergoes deprotonation to form diol 269 with the *para*-menthane framework; the second aldehyde molecule attacks diol 269 at the secondary carbon atom of the endocyclic double bond; the subsequent intramolecular cyclizations give product 267 (Scheme 48) [55]. It is important that all the transformations proceeded stereoselectively giving only one isomer of compound 267.

Similar transformations took place upon the interaction of *trans*-4-hydroxymethyl-2-carene 261 with (2*E*)-butenal 2 on clay; however, the yield of tricyclic product 272 was considerably lower than that for the reaction with aldehyde 22 [55].

After the replacement of the methyl radical with phenyl radical, when passing from aldehyde 2 to cinnamaldehyde 15, isobenzofuranes 273 and 274 exclusively were formed (Scheme 49) [55].

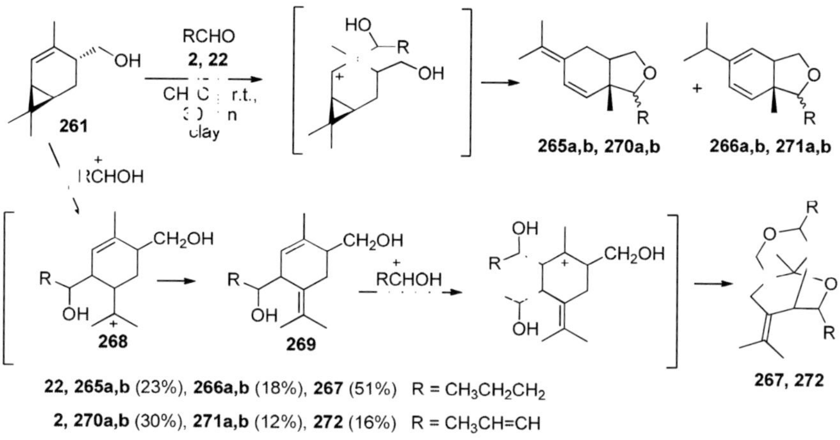

Scheme 48.

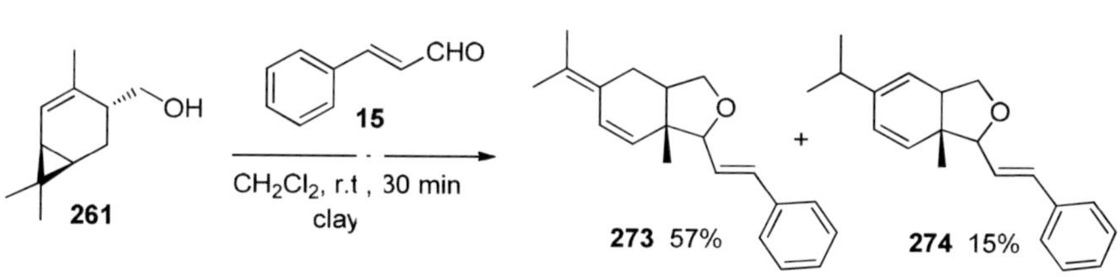

Scheme 49.

The interaction between *trans*-4-hydroxymethyl-2-carene 261 and methacrylaldehyde 93, which is isomeric to (2*E*)-butenal 2, results in a completely different course of the reaction. Due to the presence of methyl substituent at the α-position of aldehyde 93 (therefore, in intermediate 275, as well), the most favorable process is the interaction between the cation center and the terminal double bond rather than the opening of the cyclopropane ring, as it is for the reaction with (2*E*)-butenal 2 (Scheme 48). The resulting intermediate product 276 subsequently undergoes heterocyclization, protonation, and different shifts of hydrogen atoms yielding tricyclic products 277 and 278 (Scheme 50) [55].

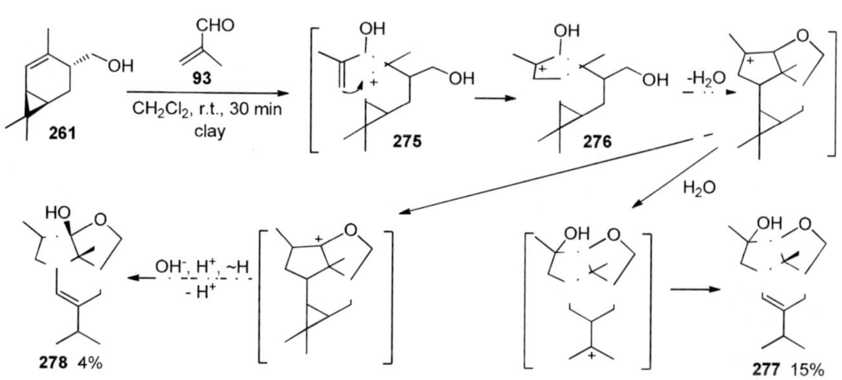

Scheme 50.

When passing to acrylaldehyde 92, tetracyclic compound 282, which presumably is the product of carbocyclization of compound 281 accompanied by binding of water, was isolated from the reaction mixture along with the diastereomeric isobenzofurans 279a,b and 280a,b and the product of binding of two aldehyde molecules compound 281 (Scheme 51) [55].

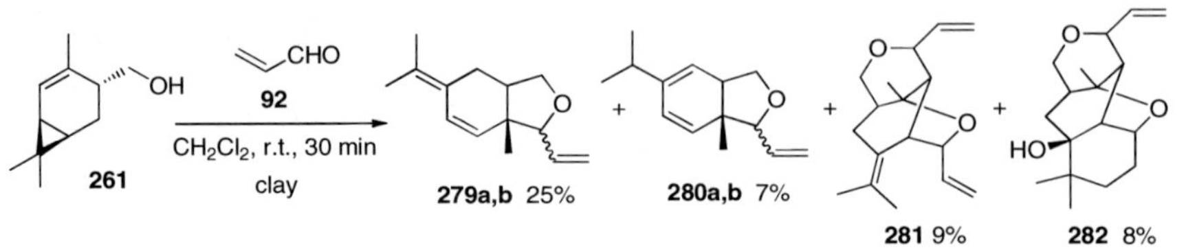

Scheme 51.

The interaction between *trans*-4-(1-hydroxyethyl)-2-carene 283 and salicylaldehyde 47 in the presence of montmorillonite clay yields tetracyclic compound 284 (20%), which is an analogue of compound 155 (Scheme 46) and bicyclic product 285 (23%) formed via intramolecular heterocyclization (Scheme 52) [6].

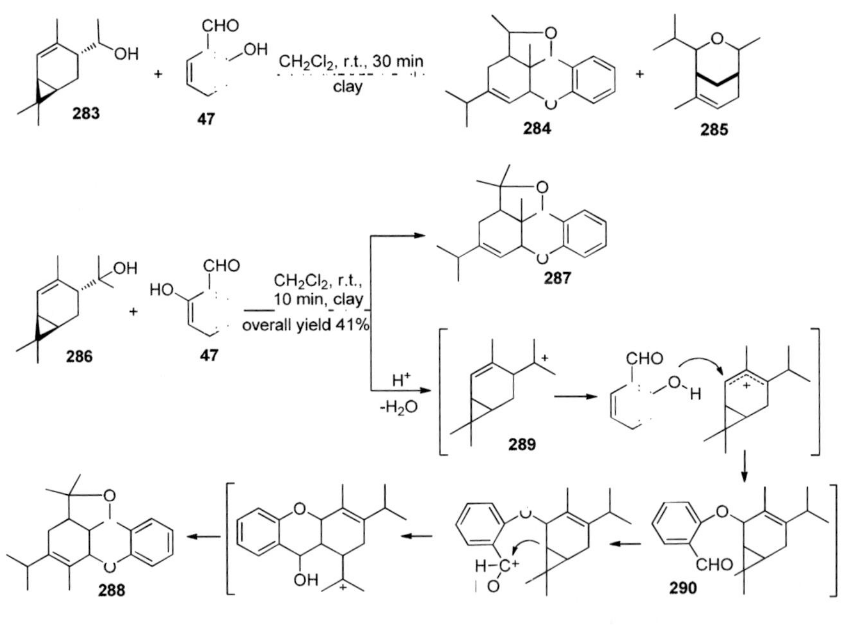

Scheme 52.

The subsequent proton detachment and tautomerization give products, such as 175 and 178 (Scheme 28). The insertion of an additional methyl group into monoterpenoid 283 when passing to compound 286 results in a considerable change in reactivity. The interaction between compound 286 and aldehyde 47 in the presence of clay yielded compound 287 (its mechanism of formation being presumably the same as that of 284) and compound 288 with another arrangement of methyl groups, assuming that the mechanism of interaction is different (Scheme 52) [6].

The formation of a relatively stable tertiary cation 289 presumably predetermines the possibility of carrying out the entire chain of transformations with aldehyde 290 being the key intermediate product. The intramolecular carbocyclization in compound 290 followed by heterocyclization gives compound 288 [6].

Sabinene 291 is a bicyclic natural monoterpene occurring in juniper essential oils. Although sabinene 291 is capable of yielding the relatively stable α-cyclopropylcarbinyl ion upon protonation of its double bond (similar to how it occurs upon protonation of 2-carene 246 and its derivatives), it is reluctant to react with aldehyde in the presence of K10 clay, unlike 2-carene 246.

Thus, sabinene 291 undergoes isomerization mostly into the compounds with the *para*-menthane framework and does not interact with 4-fluorobenzaldehyde 11, salicylaldehyde 47, 4-hydroxy-3-methoxybenzaldehyde 171 and 4-methoxybenzaldehyde 73 [56]. The intramolecular heterocyclic products 5 and 292 were isolated only for the reaction with (2*E*)-butenal 2, however, their yields were rather low as well (Scheme 53).

The formation of compound 5 presumably occurs via the repeatedly mentioned cation 6; whereas the reaction yielding compound 292 can be initiated by the attack of the protonated aldehyde at the cyclopropane ring of monoterpene 291 [56].

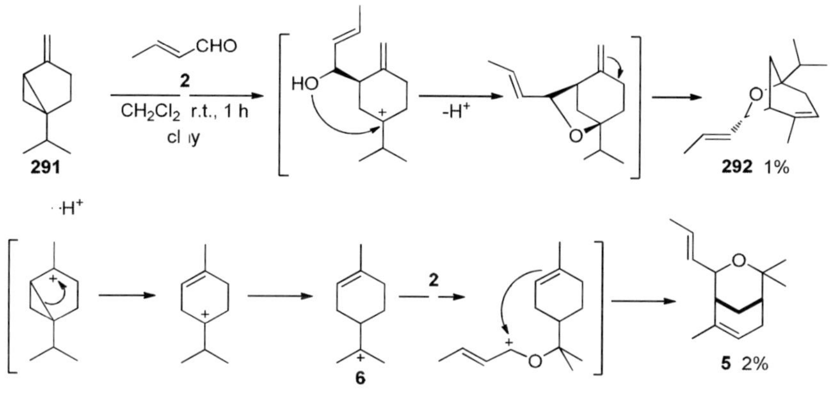

Scheme 53.

CONCLUSION

The reactions of aromatic and aliphatic aldehydes with monoterpenoids yield heterocyclic oxygen-containing compounds with various types of framework, including bi-, tri-, and tetracyclic compounds. Almost all these reactions are catalyzed by acids; the Lewis acids and montmorillonite clays being typically used as the catalysts. Some of the resulting compounds exhibited significant biological activity; a number of compounds are waiting for the attention from pharmacologists.

Due to the ability of monoterpenoids of different structural types to rearrange into carbocations with the *para*-menthane framework upon protonation, a number of identical heterocyclic compounds can be obtained using various starting terpenoids, thus providing chemists with broad opportunities for substrate selection. Meanwhile, a number of polycyclic compounds are formed from monoterpenoids having a certain structure and a certain arrangement of functional groups.

Despite the fact that most (in some cases, all) asymmetric centers within a terpenoid take part in the reactions, the transformations typically proceed stereoselectively. The availability of a number of monoterpenoids in optically pure form allows one to obtain optically active heterocyclic compounds. Often both enantiomers of the compounds can be synthesized.

The structural features of aldehydes in most cases have no key effect on the transformation pathway; however, they considerably affect the yield of the reaction products. The reactions of aldehydes with verbenol epoxide and 4-hydroxymethyl-2-carene in the presence of clays are the exceptions. Furthermore, tandem heterocyclizations frequently occur in the reactions with salicylaldehyde, involving the phenolic hydroxyl group.

In general, the most complex multistage transformations are observed when performing the reactions of oxygen-containing *para*-menthane, pinane and carane monoterpenoids with aldehydes in the presence of clays, especially if there are one or several oxygen-containing substituents in the aldehydes used. This pathway has been least well-studied; the researches in this field will enable synthesizing new complex heterocyclic compounds via a single preparatory stage.

REFERENCES

[1] Salakhutdinov, N. F.; Barkhash, V. A. *Russ. Chem. Rev.* 1997, *66*, 343-362.

[2] Il'ina, I. V.; Volcho, K. P.; Salakhutdinov, N. F. *Russ. J. Org. Chem.* 2008, *44*, 1-23.

[3] Volcho, K. P.; Salakhutdinov, N. F. *Mini-Rev. Org. Chem.* 2008, *5*, 345-354.

[4] Kaiser, R.; Lamparsky, D. *Helv. Chim. Acta* 1984, *67*, 1184-1197.

[5] Volcho, K. P.; Korchagina, D. V.; Tatarova, L. E.; Salakhutdinov, N. F.; Barkhash, V. A. *Zh. Org. Khim.* 1993, *29*, 646-647.

[6] Volcho, K. P.; Korchagina, D. V.; Gatilov, Yu. V.; Salakhutdinov, N. F.; Barkhash, V. A. *Russ. J. Org. Chem.* 1997, *33*, 607-618.

[7] Saha, P.; Gogoi, P.; Saikia, A. K. *Org. Biomol. Chem.* 2011, *9*, 4626-4634.

[8] Reddy, U. C.; Bondalapati, S.; Saikia, A.K. *J. Org.Chem.* 2009, *74*, 2605-2608.

[9] Estévez, R. E.; Oller-López, J. L.; Robles, R.; Melgarejo, C. R.; Gansäuer, A.; Cuerva, J. M.; Enrique Oltra, J. *Org. Lett.* 2006, *8*, 5433-5436.

[10] Baldwin, J. E.; Thomas, R. C.; Kruse, L. I., Silberman, L. *J. Org. Chem.* 1977, *42*, 3846-3852.

[11] Bröhmer, M. C.; Volz, N.; Bräse, S. *Synlett* 2009, 1383-1386.

[12] Subba Reddy, B. V.; Divya, B.; Swain, M.; Prabhakar Rao, T.; Yadav J. S.; Vishnu Vardhan, M. V. P. S. *Bioorg. Med. Chem. Lett.* 2012, *22*, 1995-1999.

[13] Yadav, J. S.; Subba Reddy, B. V.; Mahesh Kumar, G.; Murthy, Ch. V. S. R. *Tetrahedron Lett.* 2001, *42*, 89-91.

[14] Dobbs, A. P.; Martinović, S. *Tetrahedron Lett.* 2002, *43*, 7055-7057.

[15] Liu, F.; Loh, T.-P. *Org. Lett.* 2007, *9*, 2063-2066.

[16] Yadav, J. S.; Subba Reddy, B. V.; Narayana Kumar, G. G. K. S.; Madhusudhan Reddy, G. *Tetrahedron Lett.* 2007, *48*, 4903-4906.

[17] Yadav, J. S.; Subba Reddy, B. V.; Maity, T.; Narayana Kumar, G. G. K. S. *Tetrahedron Lett.* 2007, *48*, 7155-7159.

[18] Yadav, J. S.; Subba Reddy, B. V.; Maity, T.; Narayana Kumar, G. G. K. S. *Tetrahedron Lett.* 2007, *48*, 8874-8877.

[19] Lenardao, E. J.; Botteselle, G. V.; de Azambuja, F.; Perin, G.; Jacob, R. G. *Tetrahedron* 2007, *63*, 6671-6712.

[20] Yadav, J. S.; Subba Reddy, B. V.; Ganesh, A.V.; Narayana Kumar, G. G. K. S. *Tetrahedron Lett.* 2010, *51*, 2963–2966.

[21] Asinger, F.; Thiel, M.; Baltz, H. *Monatsh. Chem.* 1957, *88,* 464-477.

[22] Dömling, A.; Ugi, I. K.; *Tetrahedron* 1993, *49*, 9495-9500.

[23] Volcho, K. P.; Tatarova, L. E.; Korchagina, D. V.; Salakhutdinov, N. F.; Aul'chenko, I. S.; Ione, K. G.; Barkhash, V. A. *Zh. Org. Khim.* 1994, *30*, 641-652.

[24] Salakhutdinov, N. F.; Volcho, K. P.; Il'ina, I. V.; Korchagina, D. V.; Tatarova, L. E.; Barkhash, V. A. *Tetrahedron* 1998, *54*, 15619-15642.

[25] Hamann, L. G.; Hoyt Meyer, J.; Ruppar, D. A.; Marschke, K. B.; Lopez, F. J.; Allegretto, E. A.; Karanewsky, D. S. *Bioorg. Med. Chem. Lett.* 2005, *15*, 1463–1466.

[26] Volcho, K. P.; Korchagina, D. V.; Salakhutdinov, N. F.; Barkhash, V. A. *Tetrahedron Lett.* 1996, *37*, 6181-6184.

[27] Volcho, K. P.; Salakhutdinov, N. F.; Barkhash, V. A. *Zh. Org. Khim.* 1999, *35*, 1583-1584.

[28] Lombard, R.; Asfazadourian, M. *Bull. Soc. Chim. France* 1961, *7*, 1311-1316.

[29] Hudson, B. J. F.; Schmerlaib, G.; *Terahedron* 1957, *1*, 284-288.

[30] Saha, P.; Reddy, U. C.; Bondalapati, S.; Saikia, A. K. *Org. Lett.* 2010, *12*, 1824-1826.

[31] Saha, P.; Saikia, A. K. *Tetrahedron* 2012, *68*, 2261-2266.

[32] Macedo, A.; Wendler, E. P.; Dos Santos, A. A.; Zukerman-Schpector, J.; Tiekink, E.R.T. *J. Braz. Chem. Soc.* 2010, *21*, 1563-1571.

[33] Silva Jr., L. F.; Quintiliano, S. A. *Tetrahedron Lett.* 2009, *50*, 2256-2260.

[34] Bondalapati, S.; Reddy, U. C.; Saha P.; Saikia, A. K. *Org. Biomol. Chem.* 2011, *9*, 3428-3438.

[35] Gurudutt, K. N.; Pasha, M. A.; Ravindranath, B. *Indian J. Chem., Sect. B: Org. Chem. Incl. Med. Chem.* 1985, *24*, 820-823.

[36] Volcho, K. P.; Tatarova L. E.; Korchagina D. V.; Salakhutdinov, N. F.; Barkhash V. A. *Russ. J. Org. Chem.* 2000, *36*, 32-39.

[37] Il'ina, I. V.; Volcho, K. P.; Korchagina, D. V.; Salakhutdinov, N. F.; Mikhalchenko, O. S. *Helv. Chim. Acta* 2011, *94*, 502- 513.

[38] Tolstikova, T. G.; Pavlova, A. V.; Morozova, E. A.; Il'ina, I. V.; Volcho, K. P.; Salakhutdinov, N. F. Application of 4,7-dimethyl-2-(prop-1-enyl)-3,4,4a,5,8,8a-hexahydro-2*H*-chromen-4,8-diol as analgesic. RU Patent 2,418,578, December 8, 2009.

[39] Il'ina, I. V.; Korchagina, D. V.; Salakhutdinov, N. F.; Barkhash V.A. *Zh. Org. Khim.* 1999, *35*,491-492.

[40] Il'ina, I. V.; Korchagina, D. V.; Salakhutdinov, N. F.; Barkhash V.A. *Zh. Org. Khim.* 2000, *36*, 1483-1492.

[41] Il'ina, I.V.; Volcho, K.P.; Korchagina, D.V.; Barkhash, V. A.; Salakhutdinov, N.F. *Helv. Chim. Acta.* 2007, *90*, 353-368.

[42] Il'ina, I. V.; Korchagina, D. V.; Volcho, K. P.; Salakhutdinov, N. F. *Russ. J. Org. Chem.* 2010, *46*, 998-1001.

[43] Tolstikova, T. G.; Pavlova, A. V.; Morozova, E. A.; Il'ina, I. V.; Volcho, K. P.; Salakhutdinov, N. F. Derivatives of 4,4,7-trimetyl-2-phenyl-4a,5,8,8a-tetrahydro-4H-benz[1,3]dioxine-8-ol as analgesic agents. RU Patent 2,409,353, July 10, 2009.

[44] Tolstikova, T. G.; Pavlova, A. V.; Morozova, E. A.; Il'ina, I. V.; Ardashov, O. V.; Korchagina, D. V.; Volcho, K. P.; Salakhutdinov, N. F. 2-(4-hydroxy-3-methoxyphenyl)-4,7-dimethyl-3,4,4a,5,8,8a-hexahydro-2H-chromene-4,8-diol – novel analgesic agent. RU Patent 2,430,100, February 24, 2010.

[45] Il'ina, I. V.; Volcho, K. P.; Korchagina D. V.; Barkhash, V. A.; Salakhutdinov, N. F. *Helv. Chim. Acta.* 2006, *89*, 507-514.

[46] Blomquist, A. T.; Himics, R. J.; Meador, J. D. *J. Org. Chem.* 1968, *33*, 2462-2465.

[47] Isagulyants, V. I.; Pokrovskaya, I. E.; Homko, S.V.; Shilrinova, A. E.; Zakharova, N. A. *Doklady AN ArmSSR* 1969, *48*, 89-92.

[48] Tatarova, L. E.; Yarovaya, O. I.; Volcho, K. P.; Korchagina D. V.; Salakhutdinov, N. F.; Ione, K. G.; Barkhash, V. A. *Zh. Org. Khim.* 1995, *31*, 982-995.

[49] Fomenko, V. V.; Volcho, K. P.; Korchagina D. V.; Salakhutdinov, N. F.; .; Barkhash, V. A. *Russ. J. Org. Chem.* 2002, *38*, 370-373.

[50] Yarovaya, O. I.; Korchagina D. V.; Gatilov, Yu. V.; Barkhash, V. A. *Russ. J. Org. Chem.* 2002, *38*, 810-822.

[51] Acharya, S. P.; Brown, H. C. *J. Am. Chem. Soc.* 1967, *89*, 1925-1932.

[52] Rudakov, G. A. *Chemistry and technology of camphor*, Lesnaya promishlennost': Moscow, 1976, p 185.

[53] Il'ina, I. V.; Volcho, K. P.; Korchagina D. V.; Salnikov, G. E.; Genaev, A. M.; Karpova, E. V.; Salakhutdinov, N. F. *Helv. Chim. Acta* 2010, *93*, 2135-2150.

[54] Jayaraman, M.; Srirajan, V.; Deshmukh, A. R. A. S.; Bhawal B. M. *Tetrahedron* 1996, *52*, 3741-3756.

[55] Il'ina, I. V.; Volcho, K. P.; Korchagina D. V.; Salakhutdinov, N. F.; Barkhash, V. A. *Russ. J. Org. Chem.* 1999, *35*, 673-683.

[56] Il'ina, I. V.; Korchagina D. V.; Volcho, K. P.; Salakhutdinov, N. F.; Tolstikov, G. A. *Russ. J. Org. Chem.* 2010, *46*, 1002-1005.

Reviewed by Dr. Artem D. Rogachev, N.N.Vorozhtsov Novosibirsk Institute of Organic Chemistry, Novosibirsk, Russian Federation, rogachev@nioch.nsc.ru.

In: New Developments in Aldehydes Research
Editors: L. Torrioni and E. Pescasseroli

ISBN: 978-1-62417-090-4

Chapter 3

UPDATE ON ALIPHATIC ALDEHYDES IN LIPID FOODS

M. C. Pérez-Camino,* R. B.Gómez-Coca and W. Moreda#

Food Quality and Characterization Department,
Instituto de la Grasa, Consejo Superior de
Investigaciones Científicas (CSIC), Sevilla, Spain

ABSTRACT

This review provides information on the procedures and methodologies for the isolation, subsequent description and quantitative determination of aliphatic aldehydes, including the short- (C3-al to C10-al) and medium- (C12-al to C18-al) chain aldehydes and the long-chain ones (C22-al to C32-al). Aldehydes occur in numerous matrices, having both biological and commercial importance. Their physiological roles, the relevance of their presence and the incidence of their chain length are shown. Their relationship with other minor compounds such as alcohols and hydrocarbons is also reported. Special emphasis is put on the studies on the content and composition of the fractions containing the aldehydes in seed lipids and lipids from fruits and leaves. Also their presence in edible materials is commented. The different methodologies developed showed hexanal (C6-al) and nonanal (C9-al) as the most active compounds among the short-chain aldehydes, and hexacosanal (C26-al), octacosanal (C28-al) and dotriacontanal (C32-al) among the long-chain ones. In edible oils such as extra virgin olive oils, long-chain aliphatic aldehydes with even carbon-atom numbers from C22 to C30 are also present and can be isolated from the waxy fraction in quantities around a hundred mg kg^{-1} oil, being the C26 aldehyde the most abundant one in all of the studied samples.

Keywords: Aliphatic Aldehydes; Minor compounds; Analytical methodologies; Edible Oils; Wax esters

* Corresponding author: M. C. Pérez-Camino. E-mail address: mcperezcamino@ig.csic.es.

Food Quality and Characterization Department, Instituto de la Grasa, Consejo Superior de Investigaciones Científicas (CSIC), Avda. Padre García Tejero 4, 41012 Sevilla, Spain.

1. Introduction

The *wax* term encompasses a pool of different compounds which have in common their presence on the surface or cuticle part of the fruits and leaves. Studies on waxes have been conducted for more than a century (Tomae, 1913; Reina-Pinto and Yephremov, 2009). Waxes serve the plants as a barrier, since they form a protective coating against UV-radiation, fungal and bacterial pathogens, dust, pollen, and water loss. Beyond the simple role of bystander protective wall, it has been demonstrated recently the active function of some cuticle waxy compounds in the fungal pre-penetration process (Hansjakob *et al.*, 2011).

From the chemical point of view waxes or, properly speaking, wax esters are the result of the esterification of an alcohol with a fatty acid:

$$RCOOH + R'OH \longrightarrow RCOOR' + H_2O$$

In the case of aliphatic lineal alcohols (R´OH), the results are lineal waxes. Alcohols may be also terpenic molecules and sterols, resulting in terpenic and sterolic waxes, respectively.

Into the first concept of waxes, authors include all the cuticle lipids corresponding to their subject materials, such as fatty aldehydes, fatty hydrocarbons, wax esters, sterol esters and alkyl esters (Adhikari *et al.,* 2006), defining those groups of compounds as the main components of the plant waxes. Moreover, most authors agree that long-chain aldehydes are the major components in the surface lipids of many plants, seeds and leaves. Other authors also include primary alcohols and the terpenic oleanolic and malic acids in their definition of waxes since these compounds are present as major components in the cuticle part of certain fruits and leaves including those of the olive trees.

In the case of edible fats and oils the aforementioned compounds are present depending on their own solubility into the triglyceride matrices, as these are the major components in the media. Wax definition is therefore limited to the second option. Even more, only those waxes formed by esterification of major fatty acids with high molecular weigh aliphatic alcohols (alcohols with 20 to 28 C-atoms, which give rise to lineal waxes with 40 to 46 C-atoms), are considered important enough to be quantified in olive oils. According to the existing regulation virgin olive oil must not contain more than 250 mg kg^{-1} (EEC, 1991). In the case of sunflower oils wax esters can have up to 54 carbon atoms (C54-W) and are determined quantitatively in rare occasions (Baumler *et al.*, 2007; Kanya *et al.,* 2007). This definition leaves other wax compounds including aliphatic aldehydes out of any official determination or regulations.

The above-mentioned long-chain aldehydes are important components of the waxes of the surface of fruits and leaves, and they have scarce relationship with the presence of other groups of aldehydes, namely short- or medium-chain aldehydes (general formula C_nH_{2n+1}-CHO; n = 2-11). Short-chain aldehydes appear in fats, oils, and fatty foods as oxidation products, mainly through enzymatic processes. These reactions take place in the unsaturated chains of fatty acids, either free or attached to triacylglycerols moieties, and involve hydroperoxide formation. Their roles and mechanisms of formation have been studied for years (Margenat *et al.,* 2005; Grompone *et al.*, 2004)

2. Short-Chain Aliphatic Aldehydes

Saturated and low molecular weight aldehydes from propanal (C3-al) to undecanal (C11-al), also named as short-chain aliphatic aldehydes or alkanals (SCAA), belong to the lineal carbonilic group of compounds which are present ubiquitously in numerous natural fatty foods (Yokouchi *et al.,* 1990). These aldehydes are detected during the obtaining process and storage of the fatty matrices, and increase mainly due to heat treatments. All the processes of SCAA formation have the breakdown of hydroperoxides (mainly those coming from the unsaturated linoleic and linolenic acids) as common factor. In these processes short-chain volatile compounds including SCAA are formed and they are perceived as pleasant odor of green grass.

Studies on the split of the hydroperoxides are extensive and have been effectively accomplished throughout the years, giving ride to numerous papers from very different lines of research (Varlet *et al.*, 2007; Kalua *et al.*, 2007)

The point of view according to which volatile aldehydes confer pleasant green notes is not unique. The formation of the same type of compounds from oxidized or altered fats and oils yields the same types of aldehydes but in these processes they are not considered as desirable compounds.

The differences in the aroma lie in the concomitant effect of the pool of compounds generated in each reaction and in their relative concentrations. However, in the case of soybean and other legumes the high concentrations of n-hexanal is considered to be responsible for the unpleasant "grassy beany" flavor (Piccirillo *et al.*, 2005)

All the studies on aroma biogenesis consider short-chain volatile compounds to be mainly aldehydes. These aldehydes would be responsible for the specific and very characteristic odor notes. Their formation involves the action of the lipoxygenases (LOX), a pro-oxidant group of enzymes belonging to the oxidoreductase family. These enzymes catalyze the oxidation of linoleic and linolenic acids, or of their positional isomers, resulting hydroperoxides as primary oxidation products (Kalua *et al.*, 2007). One kind of oxidoreductase is the hydroperoxide lyase (HPL), which catalyzes the cleavage of hydroperoxides and forms short-chain aldehydes, mainly with six (C6-al) or nine (C9-al) carbon atoms, depending on the specificity enzyme-substrate.

Volatile aldehydes are important constituents of the flavor and aroma of many fruits and vegetables. Enzymes involved in aldehyde formation take part also in numerous processes and they are widely distributed in plants, animals, fungi and some bacteria. Thus, in the plant kingdom, the LOX participates in seed germination, fruit ripening, defense mechanisms against stress factors, etc. In mammals, the products of its catalytic activity are intermediates in processes such as arthritis.

Once fats, oils and fatty foods are obtained, they are exposed to oxidation processes with formation of compounds analogous to those produced by the enzyme oxidoreductase but without its presence. External agents such as light, oxygen and temperature induce those processes.

One of the most studied reactions in the oxidation process of oils and fatty foods is the one that begins by the action of light, in which one hydrogen atom is abstracted from a methylene carbon conjugated to a double bond. The consequence is an unstable radical formed from the main unsaturated fatty acids such as oleic and linoleic acids. The result is the formation of 8 to 13 carbon atom hydroperoxides by cleavage of the carbon-carbon bonds on each side of the alkoxy radical of the hydroperoxides, producing among others volatile aldehydes: hexanal, octanal, nonanal and decanal (Figure 1). The quantification of these aldehydes and, to be more precise, of the C6-al alkanal was proposed time ago as a formula to evaluate the quality of vegetable oils (Warner *et al.*, 1978) and cured sausages (Bello *et al.*, 1993; Martín *et al.*, 2000).

However, a great number of studies on oxidation of oils designate the aldehyde nonanal (C9-al) as primary rancidity indicator, having been proposed as oxidation marker (Vichi *et al.*, 2003) together with the hexanal/nonanal ratio (Koprivnjak *et al.*, 2005).

Oxidized oils contain alkanals and alkenals with 10 or 11 carbon atoms, which increase their concentrations with increasing heating time. The most recent studies on aldehyde formation by heating processes have demonstrated that the development rate is very slow and that it is closely related to the unsaturation of the oils.

Guillén *et al.* (2012) studied the formation of volatile compounds in heated sunflower and linseed oils determining that in both oils the main aldehydes are not the saturated but the unsaturated ones. The heated sunflower samples were rich in (E)-2-heptenal, (E)-2-octenal, and 2,4-decadienals; and the heated virgin linseed oil was the richest oil in (E)-2-propenal, (E)-2-butenal, and 2,4-heptadienals, these latter aldehydes reaching their maximum values very quickly.

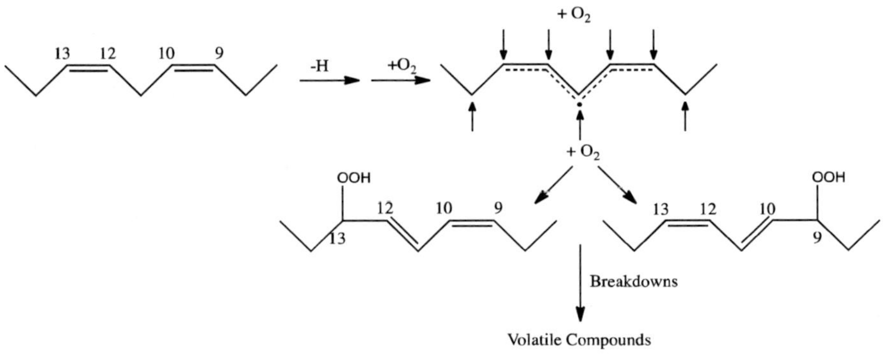

Figure 1. Oxidation process of oils and fatty foods. The reaction begins by the action of light, abstracting one hydrogen atom from a methylene carbon conjugated to a double bond, and resulting in the formation of a radical. After cleavage, 8 to 14 carbon atom hydroperoxides appear, producing among others volatile aldehydes: hexanal, octanal, nonanal and decanal.

Nevertheless Fujisaki and co-workers carried out studies on high-oleic safflower oils and determined high levels of alkanals during heating time. Figure 2 shows the evolution of the SCAA propanal, butanal, pentanal, hexanal, heptanal, octanal, nonanal and decanal, generated in high-oleic safflower oil by heating at 180°C during 30 hours in oxygen atmosphere. Although all of the SCAA increase with time, the alkanals C6-al and C9-al progress faster, reaching quantities of 7 $\mu g\ g^{-1}$ (Fujisaki *et al.*, 2002).

Studies on the composition of volatile components of Iberian hams and particularly on the occurrence of SCAAs are numerous (Martín *et al.*, 2000; Jurado *et al.*, 2009; García-González *et al.*, 2009; Narváez *et al.*, 2010). Thus, a number of research groups have accomplished studies on volatile evolution during processing, the effect of the feeding system, the influence of the anatomical location, etc. (Martín *et al.*, 2000). They have determined that SCAA from C4-al to C10-al help to strengthen and intensify the aroma in dry-cured ham and, although alkanals are recognized to present a low odor threshold they are considered to be largely responsible for the whole of the aroma produced. In these studies also the aldehydes C6-al and C9-al are the largest components, showing quantities at the end of the conservation period higher than 50000 ng g^{-1} and 2000 ng g^{-1} of dried matter for C6-al and C9-al, respectively.

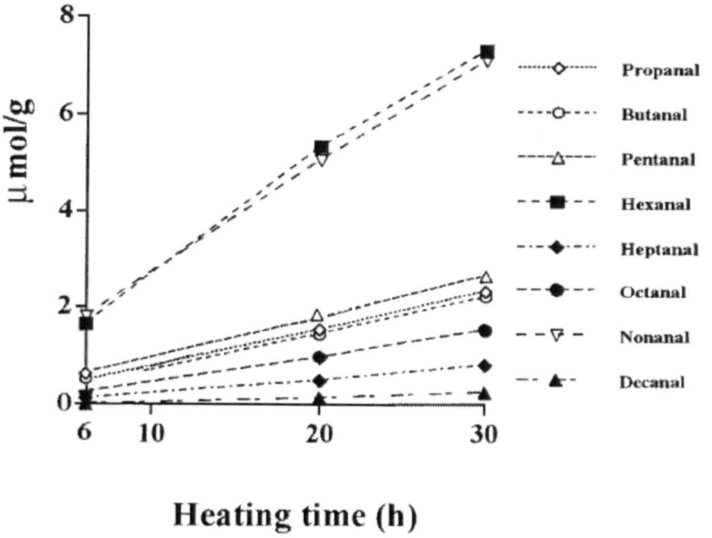

Adapted from Fujisaki *et al.*, 2002.

Figure 2. Alkanals generated in high-oleic safflower oil. The oil was heated and water-sprayed at 180°C during 30 hours in atmospheres with different oxygen levels.

Jurado *et al.* (2009) determined the effects of the feeding system and the ripening time, establishing that this last parameter affects more than the first one to the volatiles. Actually the feeding system has no effect on the SCAAs (C6-al to C9-al) at all. The aldehyde C6-al was one of the most abundant volatile components at the end of the post-salting period (120 days) and in the finished product (722 days).

The influence of the anatomical locations of the samples in Iberian dried-cured ham on the volatile compounds was studied by Narváez *et al.* (2010). In their studies the SCAA detected were from C5-al to C9-al, being pentanal the most abundant aldehyde in the adipose tissue and also in the muscle.

Recently, studies on the flavor of pomegranate fruit using solvent-assisted flavor evaporation (SAFE) and headspace solid-phase micro-extraction (HS-SPME) methods have been undertaken. Authors found the aldehyde C6-al among the most abundant volatile compounds. The alkanals C8-al and C9-al could be determined, but using GC–MS analysis of the HS-SPME extracts instead of SAFE. (Melgarejo *et al.,* 2011; Mayuoni-Kirshinbaum *et al.*, 2012; Calín-Sánchez *et al.*, 2011)

Short-Chain Aldehyde Activities

It has been demonstrated that short-chain aldehydes present activities against different fungal and bacterial strains. Noteworthy is the activity of some of the volatile flavor components of olive fruits and oils against different fungal and bacterial strains. Thus, Kubo and coworkers (1995) demonstrated the antimicrobial activity of saturated and unsaturated aldehydes from olive fruit against a broad spectrum of food-borne microfungal and bacteria strains. Also Bisignano *et al.* (2001) studied the activity of some of these compounds against a number of standard bacteria strains that could be causal agents of human infections. More recent studies have shown the antifungal activity of SCA from olive fruit against Tricophyton mentagrophytes and Microsporum. This reported antifungal activity of the aldehydes from olive fruit substantiates the use of olive and olive oil in skin diseases and suggests that these natural compounds could be useful agents in the topical treatment of the named infections (Battinelli *et al.*, 2006)

In relation to the most active aldehydes, C6-al and C9-al are assumed to be the main active components that inhibited the growth of the fungus. Nevertheless, the unsaturated series is more active than the saturated one. This series has proven its effectiveness against all dermatophytes tested. Particularly (E)-2-hexenal and (E)-2-nonenal showed a broader spectrum than the corresponding saturated compounds (Battinelli *et al.*, 2006). These data confirm previous results obtained by Bisignano *et al.* (1999) who asserted that the presence of a double bond in the molecule was important for the antimicrobial activity of olive aldehydes.

Battinelli *et al.* (2006) reported data on the 'in vitro' activities of eight SCAA from olive flavor: hexanal, nonanal, (E)-2-hexenal, (E)-2-heptenal, (E)-2-octenal, (E)-2-nonenal, (E)-2-decenal and (E,E)-2,4-decadienal, against a number of standard and freshly isolated bacterial strains that may be causal agents of human intestinal and respiratory tract infections. In this particular study the saturated aldehydes did not exhibit significant antibacterial activity, but the unsaturated ones showed a broad antimicrobial activity spectrum. It is concluded that the effectiveness of the aldehydes seems to depend not only on the presence of a double bond, but also on the chain length from the *enal* group and on the microorganism tested.

The ability of the aldehydes to produce fluorescent compounds with similar properties (same maximum excitation/emission) as the ones obtained during fish processing was tested. Among the SCAA the alkanals C3-al and C6-al were reacted in model systems with sardine *(Sardina pilchardus)* muscle and with n-propylamine as a source of nucleophilic compounds, at 30°C in dark during 25 days. A fluorescence shift was detected for the aldehydes investigated as long as the reaction time increased. Great differences were observed according to their chemical structure and reactivity (Aubourg, 1998).

Volatile Aldehyde Isolation and Quantitative Determination

Diverse techniques can be used for volatile compound analysis and so for aldehyde extractions. Solid-phase micro extraction (SPME) and dynamic headspace/static headspace (DHS/SHS) are the methods recently utilized with the aim of isolating volatile compounds from matrices such as seafood -salmon, crayfish, sardine prawns, etc- and also from ham. All these studies have been addressed in excellent and comprehensive papers (Varlet *et al.*, 2007)

Liquid–liquid extractions (LLE) have also been used but are not very suitable to recover volatile aldehydes. However, recent devices for LLE like supercritical fluid extraction (SFE) (Aro, *et al.*, 2002) or microwave-assisted extraction (MAE) (Grimm *et al.*, 2000) are more and more used. Descriptions of the volatiles and also quantitative results are presented when an appropriated internal standard is included in the analytical process.

Beránek and Kubátová (2008) developed and optimized a method for the trace analysis of a wide range of aldehydes. They compared the efficiency of headspace (HS) aldehyde extraction with their on-fiber derivatization (HS-SPME-OFD) as well as that of direct derivatization (DD) in aqueous solutions followed by SPME from either headspace (DD-HS-SPME) or liquid-phase (DD-L-SPME). Their optimized method consisted of the extraction of derivatized aldehydes by liquid-phase SPME (DD-LP-SPME) at 80 °C for 30 min. The limits of detection (LODs) using this optimal method were in the range of 0.1–4.4 μg L^{-1} for the majority of SCAA. Headspace SPME-OFD using on-fiber derivatization generally showed a lower sensitivity and several compounds were not detected.

The optimized headspace SPME of derivatized aldehydes in aqueous solution was not as sensitive as DD-LP-SPME for hydroxylated aromatic compounds. The developed method was used to analyze aqueous particulate matter extracts; this method achieved higher sensitivities than those obtained with US Environmental Protection Agency (EPA) Method 556 (Munch *et al.,* 1998) Guillén and Uriarte (2012) studied by SPME-GC–MS relationships between the abundances of the volatile aldehydes and the percentages of altered or polar compounds in three edible oils. Among the studied volatile compounds the alkanals C6-al and C9-al were the most abundant. Automated head-space solid-phase micro-extraction (A-HS-SPME) coupled to gas chromatography–ion trap mass spectrometry (GC–ITMS), was also developed and employed for fast characterization of olive oil volatiles (Cajka *et al.*, 2010) Figure 3 shows a A-HS-SPME–GC–ITMS chromatogram of olive oil volatiles. The separations have been done with a combination of two GC columns: HPINNOWax (polyethylene glycol) and BPX-50 (50% phenyl polysilphenylenesiloxane) Propanal, pentanal, hexanal, octanal and nonanal are the main SCAA determined in those conditions. Figure 4 corresponds to a gas-chromatogram of volatiles of virgin olive oil carried out on a polar chromatographic capillary column (Supelcowax-10), and a DVB–CAR–PDMS SPE-fiber. The retention times of aldehydes are shorter than those of alcohols and acids of the same chain length; thus, hexanal elutes before hexanol and hexanoic acid (40 °C, 10 min; 3 °C min^{-1} up to 200 °C). Different SPME fibers for volatile extraction have been tested: PDMS 100 T μm, CAR–PDMS 75 μm, PDMS–DVB 65 μm, and DVB–CAR–PDMS 50 and 30 μm, 2 cm long (Vicki *et al.*, 2003)

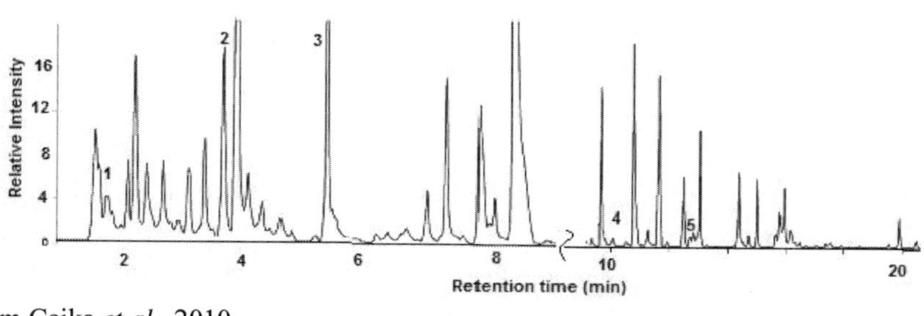

Adapted from Cajka *et al.,* 2010.

Figure 3. HS-SPME–GC–ITMS chromatogram of olive oil SCAA. The separations have been done with a combination of two GC columns: HPINNOWax (polyethylene glycol) and BPX-50 (50% phenyl polysilphenylenesiloxane); 1: propanal; 2: pentanal; 3: hexanal; 4: octanal; 5: nonanal.

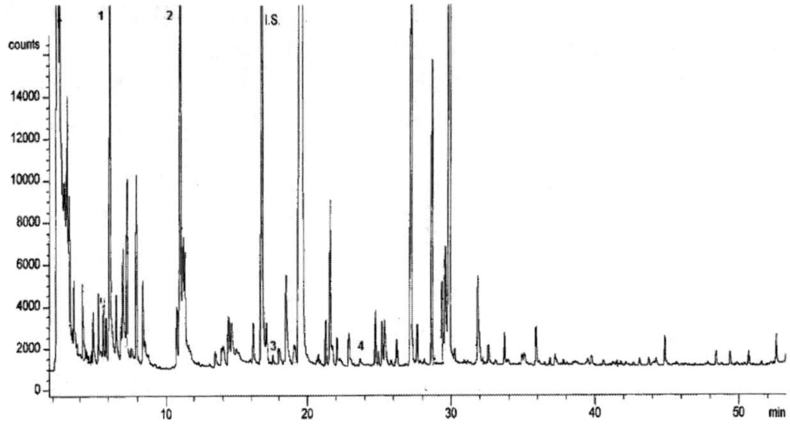

Adapted from Vichi *et al.,* 2003.

Figure 4. Gas chromatogram of volatile SCAAs of virgin olive oil carried out on a polar chromatographic capillary column (Supelcowax-10), and a DVB–CAR–PDMS SPE-fiber;1: pentanal; 2: hexanal; 3:heptanal; 4: octanal.

Recent studies on sensory analysis and volatiles of Egyptian virgin olive oils using SPME-GC-MS have been carried out by Benincasa *et al.* (2011). They have shown the volatile aldehydes heptanal (C7-al) and octanal (C8-al) as the major saturated ones with quantities up to 254 mg kg^{-1} for C7-al and up to 570 mg kg^{-1} for C8-al, although there were other analysis where those aldehydes were not detected. These studies have been done using a GC capillary column VF-5ms 60 m × 0.25 mm x 0.25 μm. The quantitative results were performed using the alcohol 2-methyl-4-pentanol as internal standard. The volatile fraction extraction was done with a DVB-CARB-PDNS 70 SPE-fiber. The temperature program of the GC was slightly different than in other studies: 50 ºC 10 min, then increase up to 180 ºC at 25 ºC min^{-1}, and then ramp to 220 °C at 10 °C min^{-1}.

2. Medium-Chain Aldehydes

Medium-chain alkanals (MCAA) comprise those aliphatic aldehydes from C12-al to C17-al and are considered intermediates compounds or by-products in the α-oxidation of long-chain fatty acids. One of the first papers reporting the presence of MCAA goes back to the 70´s with the work of Galliard and Mattew (1976). These authors reported the biogenesis of penta-, tetra-, tri- and dodecanal (C12-al to C15-al), all of them present in cucumber fruits (*Cucumis sativus*). The formation process of these MCAA in the extracts of cucumber involve the α-oxidation of palmitic acid. The α-oxidation of high molecular weight fatty acids is carried out by mediation of an enzyme system, which catalyzes the hydroxylation at carbon 2 (Figure 5). The oxidation process in plants was already elucidated in Stumpf research group of work at the end of the 50´s (Martin *et al.,* (1959). The process involved the oxidation of a C_n fatty acid to the Cn-1 aldehyde and the subsequent transformation into the C_{n-1} acid, which continues the route. The result of these linked reactions is normally an oxidative decarboxylation with formation of one-carbon-less fatty acids as final products. The first stage on the α-oxidation reaction is a mixed-function oxidation and the second one an oxidative decarboxylation. As the process goes on other aldehydes will result from the sequential removal of carbons from the carboxyl end of the fatty acid and C14-al, C13-al and C12-al are also reported. Those aldehydes are present in minor concentrations compare with those found for C15-al. Therefore, MCAA are present in cucumbers according to the series: C15-al > C14-al > Cl3-al > C12-al.

The aldehydes pentadecanal (C15-al) and heptadecanal (C17-al) occur in a wide range of green, brown and red seaweeds, and have been identified as characteristic compounds in species of the Green marine algae *Ulvaceae* (Kajiwara *et al.*, 1987). Their determination was made by GC-FID using a DB-1 fused silica capillary column of 60 m x 0.25 mm. x 0.25 μm. The column temperature was kept at 50 °C for 5 min and increased at 3 °C min^{-1} up to 240 °C.

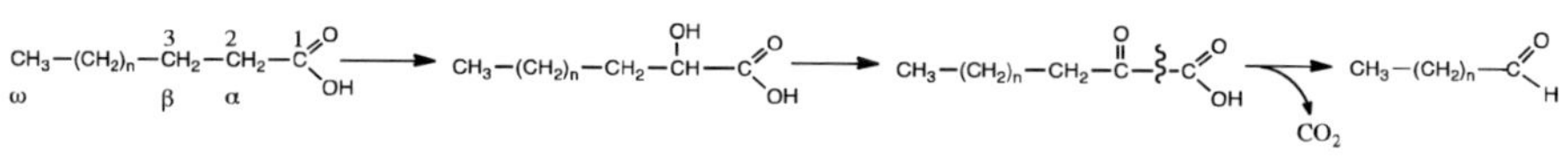

Figure 5. α-Oxidation of high molecular weight fatty acids. An enzyme system catalyzes the hydroxylation at carbon 2 and the formation of the corresponding Cn-1 acid.

In spite of all this important research work the mechanism of α-oxidation of fatty acids to produce MCAA is still poorly understood.

3. LONG-CHAIN ALDEHYDES

Straight and long-chain aliphatic aldehydes (LCAA) from 22 to 32 carbon atoms (C22-al to C32-al) are known as one of the lipid biomarker in leaves and roots of plants (Jansen *et al.*, 2006). They were reported to come from very long-chain fatty acids (LCFA) with even carbon number, which suffer an α-oxidation identical to that described for medium-chain aldehydes. Firstly the fatty acids are transformed into n-aldehydes with one carbon-atom less than originally; subsequently there is a decarboxylation process to convert the n-aldehydes into their corresponding n-alkanes, alcohols, and wax esters (Iyer *et al.*, 1998; Riendeau *et al.*, 1985; Galliard *et al.*, 1976)

In relation to the abundance of the different alkanals, the aldehyde octacosanal (C28-al) has been described as the major component of wheat epicuticular and also as its main active compound (Reisige *et al.*, 2006; Morris *et al.*, 2000)

High concentrations (55–60%) of LCAAs have been found in the waxy fraction of sugercane. Asikin and coworkers (2012) studied the sugarcane wax composition by HPLC coupled to evaporative light scattering detector (ELSD), and determined that the sugarcane waxes were comprised mainly of aldehydes. The highest content of LCAA, 600 mg 100 $g-^{1}$ rind, was found in hand-peeled rinds of specific sugarcane cultivars (Figure 6). The content of these compounds increased up to 72% during sugarcane maturation from October to January. These studies indicated that the composition and content of LCAA vary depending on the cultivar of the sugarcane and on the specific part of the sugarcane analyzed, as well as on the maturity.

Studies undertaken on LCAA in olive oils have shown the even series from docosanal to triacontanal (C22-al to C30-al) as the main one, with clear predominance of hexacosanal (C26-al), followed by octacosanal (C28-al) and tetracosanal (C24-al). The values ranged from

23.3 to 81.7 mg kg^{-1} oil. The quantitative evaluation was carried out using the on-house synthesized C21-al as internal, although the available C32 wax could be used as internal standard too (Pérez-Camino *et al.*, 2003)

Specifics maize lines have been found to greatly exhibit elevated levels of very-long-chain aldehydes, among which n-dotriacontanal (C32-al) accounted for up to 60% of the wax constituents. The significant role that LCAA plays in the initiation and promotion of the fungal pre-penetration processes in vivo on the surface of a living plant leaf, was demonstrated by Hansjakob *et al.* (2011). In a mutant which is completely devoid of cuticle very-long-chain aldehydes, germination and appressorial differentiation of B. *graminis* were strongly impeded (Avato *et al.*, 1985). The aldehyde-deficient mutant line showed a dramatic reduction of conidial germination and differentiation when compared to aldehyde-containing leaf surfaces of other maize lines. Nevertheless, after spraying the mutant leaf surface with n-hexacosanal the fungal pre-penetration process was fully restored allowing germination completely. The results of the in-vivo and in-vitro inoculation experiments corroborate unequivocally the importance of very-long-chain aldehydes for the successful initiation of the barley powdery mildew pre-penetration processes.

Research on grapefruits has shown concentrations of 35-46% of aldehydes in wax cuticle and has evidenced that cold storage affects the amount of epicuticular wax, particularly concerning to LCAA. Thus, LCAA are lost when the fruit is stored for 7 days indicating that the fruit's natural barrier to water loss and harmful gases has deteriorated. The main aldehydes belong to the even series with predominance of docotriacontanal (C32-al). At 5 ºC, when the fruit was de-waxed it suffered important chilling injuries compared with control samples, demonstrating the important role of the wax components as quality protectors. Other lines of research have lead to conclude that the synthesis of wax components could be turned on or off by moving the fruit back and forth between 5 and 15 ºC (Nordby *et al.*, 1990 and 1991; McDonald *et al.*, 1993)

Analysis of Long-Chain Aldehydes

Gas Chromatography Studies

Analysis of long-chain fatty aldehydes by GLC can be done on both polar and non-polar liquid phases giving in both cases quantitative and reproducible values. The aldehydes are eluted at relatively low temperatures on both phases and are, in general, well resolved without overlapping from other components as alkyl esters and wax esters. Nevertheless, they are not well resolved from the unsaturated hydrocarbons present in olive oils on a SPB-5 column. In our studies on olive oils, it is necessary to do a previous separation of the oil's interfering compounds on a Si-SPE column. The column is washed with 7 mL of an hexane:toluene (85:15, v/v) blend in order to eliminate the unsaturated hydrocarbons (mainly squalene) which interfere in the quantitation of LCAA on the GC. Aldehydes and wax esters are subsequently eluted from the SPE cartridge with 10 mL of the same admixture and determined quantitatively by on-column gas chromatography (Pérez-Camino *et al.,* 2012) using an SPB-5 capillary column (12 m x 0.25 mm x 0.25 μm) coated with 5% dimethyl–95% diphenylpolysiloxane. The chromatographic conditions were: oven temperature, 50 ºC for 2 min and then increased at 20 ºC min^{-1} up to 160 ºC; again increased at 4 ºC min^{-1} up to 350 ºC. The injector and detector temperatures were 50 ºC and 365 ºC, respectively.

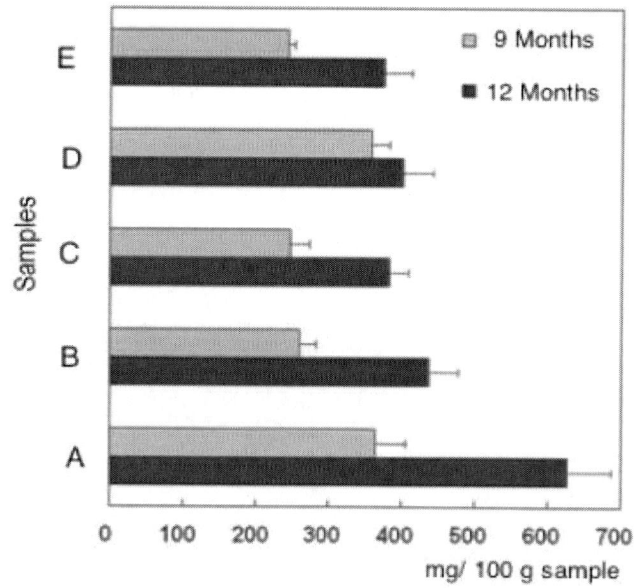

Adapted from Asikin *et al.,* 2012.

Figure 6. Data of long-chain aldehyde from immature (9 months) and mature (12 months) hand-peeled sugarcane rind samples, obtained using GC-FID.

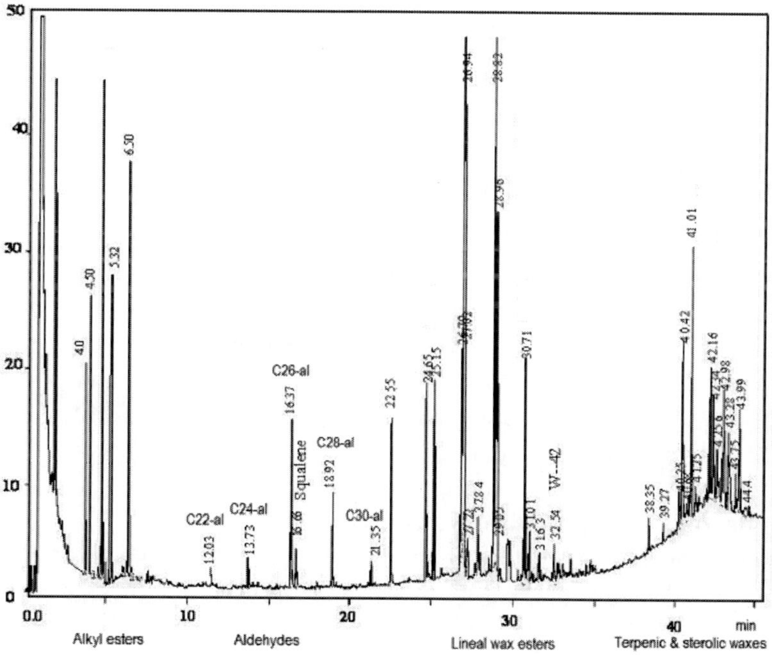

Figure 7. On-column gas chromatogram of the wax fraction of olive oil contaning alkyl esters, lineal terpenic aldehydes and waxes.

Under those conditions we find that the retention times of the aldehydes present in olive oils from dodecanal to triacontanal (C22-al to C30-al) are higher than the alkyl (methyl and ethyl) esters present in the oils: methyl palmitate to ethyl linoleate, and lower than the wax ester with their same carbon number (Figure 7). As it has been already mentioned, C26-al was the most abundant aldehyde that could be detected.

At the end of the 60´s it was already evidenced that in gas chromatography the retention times of MCAA were shorter than those of methyl esters of the same chain length, following the same pattern than SCAAs. Thus, olealdehyde (C18-al) elutes with methyl stearate on a EGSS-X liquid phase column operating at 130 to 180 °C at 3 °C min^{-1} as can be observed from Figure 8 (Wood *et al.*, 1969)

In these studies it was also clear that aldehydes eluted unchanged under the preparative conditions. In the same way different authors have observed no detectable qualitative or quantitative change during TLC or GLC of the aldehyde mixtures, despite a common belief that free aldehydes are unstable compounds (Berdyshev, 2011)

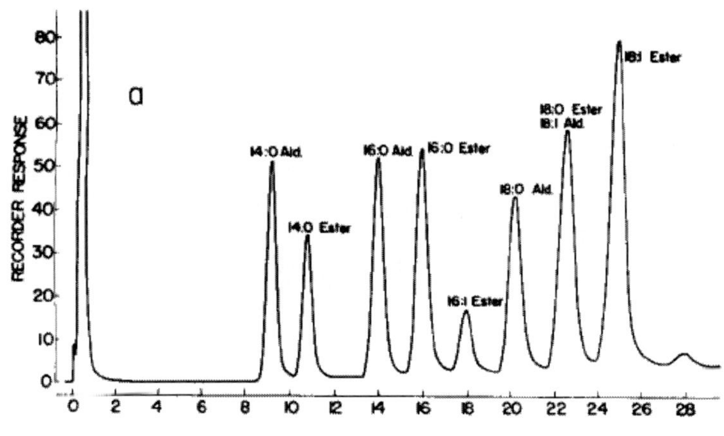

Adapted from Wood and Harlow, 1969.

Figure 8. Gas chromatogram of aldehydes and fatty acid methyl esters with the same chain length.

High Performance Liquid Chromatography

Although it is not as usual as the determinations by GC, the waxy materials from perilla, sesame seeds and sugarcane have been analyzed by high performance liquid chromatography (HPLC). The separations were done using a normal-phase silica column (250 mm x 4.6 mm x 5 µm) and evaporative light scattering (ELSD) as detector, operating at 40 °C with nitrogen pressure of 3.5 bars and a flow of 1 mL min^{-1}. A binary gradient solvent of hexane as solvent A and 0.2% acetic acid in methyl *tert*-butyl ether as solvent B was used.

In those conditions total aldehydes elute as a single peak but joined the wax esters and constitute almost 100% in the sesame seed whilst, whereas in the perilla seed are about 50%.

A representative HPLC chromatogram of the waxy material extracted from perilla seed is shown in Figure 9, where it can be observed that aldehydes are well separated from hydrocarbons, triglycerides, acids and policosanols, but not from wax esters. The LOD of this analysis was 0.01 mg mL^{-1} (Asikin *et al.*, 2012)

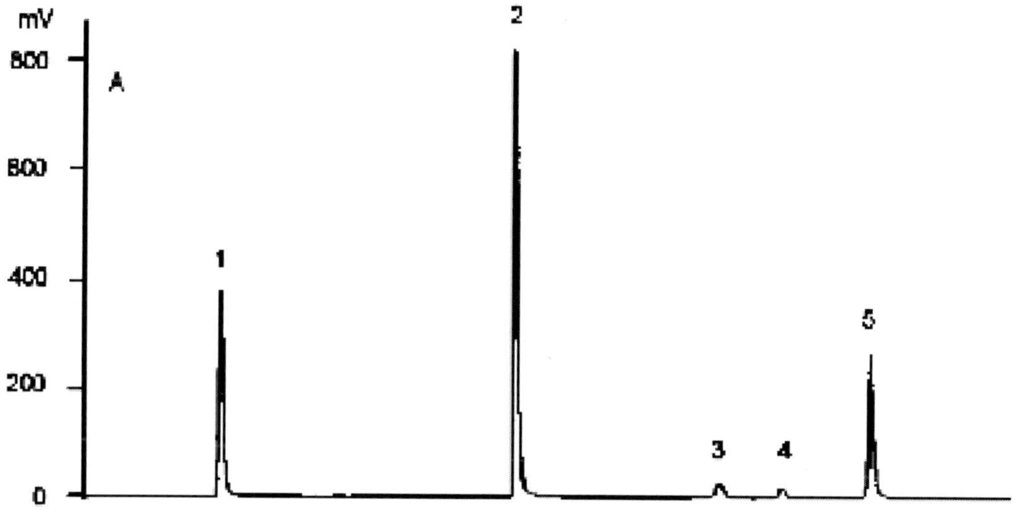

Adapted from Adhikari *et al.,* 2006.

Figure 9. HPLC of 1: hydrocarbon, 2: wax esters, 3: triglycerides, 4: acids, and 5: policosanols.

Studies of LCAA by TLC

Thin layer chromatography (TLC) has been widely used for separation and qualitative analysis of the major classes of wax components (Hwang *et al.,* 2002; Adhikari *et al,.* 2006; Webster *et al.,* 2006). The most common solvent is a medium-polar one, usually the admixture hexane:diethyl ether:acetic acid. Their proportions vary and also one or two steps of developing solvents can be carried out to elute samples.

The different bands corresponding to the wax components isolated from grain sorghum, perilla, and sesame seeds (Figure 10A), are well separated using one step of the development solvent: hexane:diethyl ether:acetic acid (85:15:2, v/v/v).

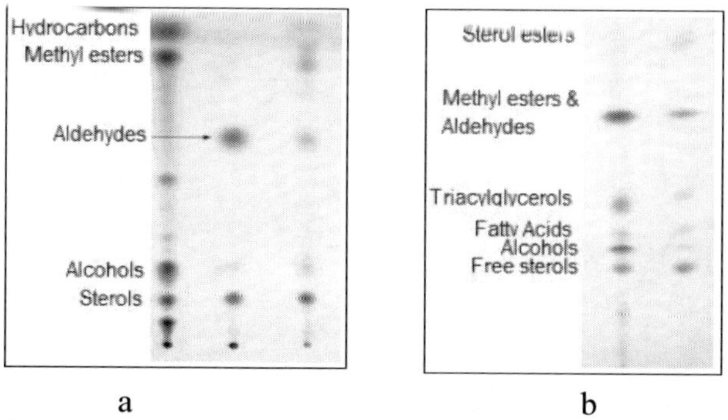

[a]Adapted from Adhikari *et al.,* 2006.
[b]Adapted from Asikin *et al.,* 2012.

Figure 10. Silica gel 60 thin-layer chromatography of waxy materials. A) Extracted from grain sorghum, perilla, and sesame seeds using hexane:diethyl ether:acetic acid (85:15:2, v/v/v) as developing solvent. (B) Extract from sugarcane using two step development solvent: hexane:diethyl ether:acetic acid 95:5:1, v/v/v, till 10 cm up the plate, and subsequently 80:20:1, v/v/v.

Hydrocarbons, methyl esters, aldehydes, alcohols, and free sterols could be made visible by spraying the plate with a 10% cupric sulfate solution containing 8% phosphoric acid. Then the TLC plate was dried until the developed bands were charred.

Figure 10B shows a two-step development process. The mixture hexane:diethyl ether:acetic acid is first applied with the proportions 95:5:1, v/v/v, and subsequently at 80:20:1, v/v/v. These mixtures have been used to isolate the components of the waxy materials extracted from sugarcane (Asikin *et al.,* 2012).

The wax components are well charred and separated, including the properly wax esters which split into two bands identified by Adhikari *et al.* (2006) methyl esters and aldehydes, and sterol esters. Methyl esters and aldehydes appear as a single spot, while fatty acids and free alcohols can be seen as close dense spots.

Studies on the Identification of LCAA by GC-MS

The LCAAs series can be undoubtedly identified using their chemical ionization (CI) and electron-impact (EI) mass spectrometry. In the analysis of the even series from C22-al to C30-al carried out without derivatization, the molecular ion [M] peaks at m/z 324, 352, 380, 408 and 436, although small were recognizable.

The characteristic losses from the molecular ions are the protonated molecules $[M-1]^+$ corresponding to the loss of H, and those losses characteristics of the aldehydes that distinguish them from alcohols: $[M-18]^+$ (loss of H_2O), $[M-28]^+$ (loss of CO), and $[M-44]^+$ (loss of CH_3CHO) (Stephanou, 1989).

The m/z values displayed in Table 1 correspond to characteristic fragments of LCAAs from C22-al to C30-al (Pérez-Camino *et al.*, 2003; Berdyshev, 2011).

Table 1. Mass spectra of long-chain aldehydes

Aldehyde	Target ion (m/z)	Qualifier ions (m/z)
Docosanal (C22-al)	324	278, 280, 306
Tetracosanal (C24-al)	352	306, 308, 334
Hexacosanal (C26-al)	380	334, 336, 362
Octacosanal (C28-al)	408	362, 364, 390
Triacontanal (C30-al)	436	390, 392, 418

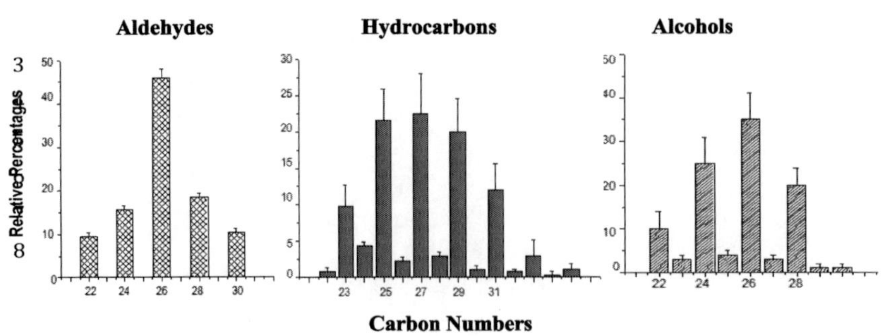

Adapted from Pérez-Camino *et al.,* 2012.

Figure 11. Relative composition (%) of aldehydes, hydrocarbons and alcohols from virgin olive oil.

Studies on the Relationship of LCAA and Other Minor Compounds

It has been described that long chain fatty acids (LCFA) with even C-atom number suffer an α-oxidation being firstly transformed into n-aldehydes with one carbon atom less. Subsequently, a long-chain-aldehyde dehydrogenase catalyzes the decarboxylation of the LCAAs to alkanes (Iyer, *et al.,* 1998; Riendeau *et al.*, 1985; Galliard *et al.*, 1976). At the same time, it has been demonstrated the existence of an aldehyde reductase that transforms aldehydes into alcohols. Those facts fit with reported results, since in olive oils studies the majority of hydrocarbons coming from some specific olive varieties (Guinda *et al.*, 1996; Koprivnjak *et al.,* 2005) are those with 25 carbon atoms, which could come from the abundant C26-al. Figures 11 and 12 show the major aldehydes, hydrocarbons, and alcohols from virgin olive oil and maize leaves, respectively. Figure 13 shows the levels of hydrocarbons determined in fresh grapefruit compared with LCAAs. In all cases the main hydrocarbons belong to the odd series. The C25 and C27 hydrocarbons are predominant in olive oils, whereas in grapefruits, C29 and C31 hydrocarbons were mainly quantified.

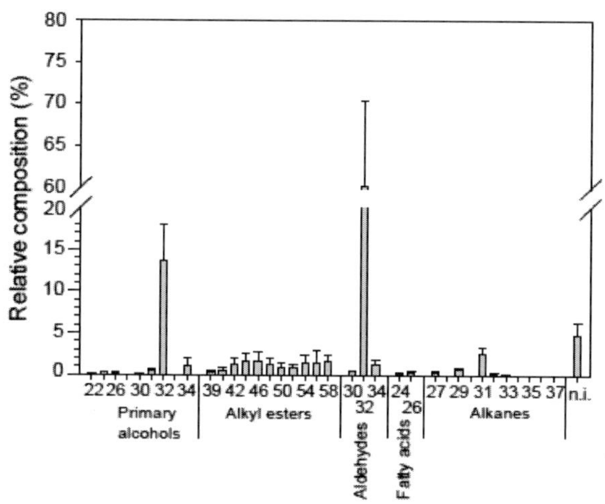

Adapted from Hansjakob *et al.*, 2011.

Figure 12. Relative composition (%) of total leaf wax from an speciphic Zea mays line.

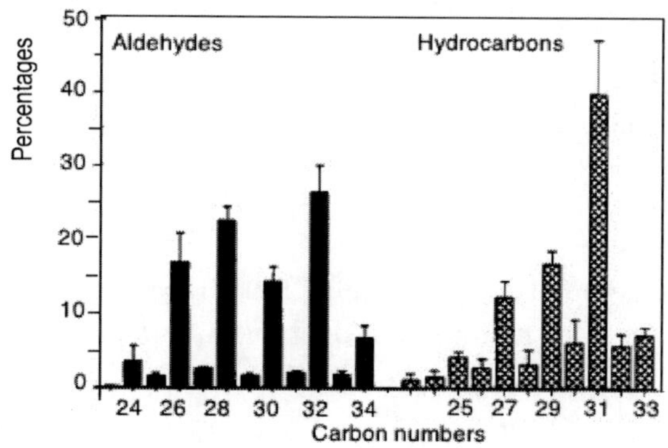

Adapted from Nordby and McDonald, 1991.

Figure 13. Relative composition (%) of aldehydes and hydrocarbons from fresh grapefruits.

This is in agreement with the fact that the major aldehydes are those from the even series, with C26-al as the most abundant one in olive and C32-al in grapefruits. Finally, data on alcohol distribution match with data on aliphatic aldehydes where also the C26 and C32 compounds are the most abundant ones in olive oil and maize leaves, respectively.

REFERENCES

Adhikari, P., Hwang, K. T., Park, J. N., and Kim, Ch. K. 2006. Policosanol content and composition in Perilla Seeds. *Journal of Agricultural and Food Chemistry* 54, 5359-5362.

Aro, T., Brede, C., Manninen, P., and Kallio, H. 2002. Determination of Semivolatile Compounds in Baltic Herring (Clupea harengus membras) by Supercritical Fluid Extraction-Supercritical Fluid Chromatography-Gas Chromatography-Mass Spectrometry. *J. Agric. Food Chem.* 50, 1970– 1975.

Asikin, Y., Takahashi, M., Hirose, N., Hou, De-X., Takara, K., and Wada, K. 2012. Wax, policosanol, and long-chain aldehydes of different sugarcane (*Saccharum officinarum* L.) cultivars. *European J. Lipid Sci. Technol.* 114, 583-591.

Aubourg, S. 1998. Fluorescence detection in aldehyde containing model systems: Relationship with fish deterioration. *Grasas Aceites* 49, 419-424

Avato, P., Bunchi, G., and Salamini, F. 1985. Absence of long chain aldehydes in the wax of the glossy ii mutant of maize. *Phytochem.* 24, 1995-1997.

Battinelli, L., Daniele, C., Cristiani, M., Bisignano, G., Saija, A., and Mazzanti, G. 2006. In vitro antifungal and anti-elastase activity of some aliphatic aldehydes from *Olea europaea* L. Fruit. *Phytomed.* 13, 558–563.

Baumler, E. R., Crapiste, G. H., and Carelli, A. A. 2007. Sunflower-oil wax reduction by seed solvent washing. *J. Am. Oil Chem. Soc.* 84, 603-608.

Bello, J., Chasco, J., Beriaín, M. J. 1993. Volatile aldehydes and lipid oxidative indices in "chorizo de Pamplona". *Revista española de ciencia y tecnología de alimentos* 33, 319-332.

Benincasa, C., Russo, A., Romano, E., Elsorady, M. E., Perri, E., and Muzzalupo, I. 2011. Chemical and sensory analysis of some Egyptian virgin olive oils. *J. Nutr. Food Sci.* 1:118. doi:10.4172/2155-9600.1000118.

Beránek, J., and Kubátová, A. 2008. Evaluation of solid-phase microextraction methods for determination of trace concentration aldehydes in aqueous solution *J. Chromatogr.* A. 31, 1209, 44-54.

Berdyshev, E. V. 2011. Review. Mass spectrometry of fatty aldehydes. *Biochim. Biophysic Acta* 1811, 680–693.

Bisignano, G., Lagana, M. G., Trombetta, D., Arena, S., Nostro, A., Uccella, N., Mazzanti, G., and Saija, A. 2001. In vitro antibacterial activity of some aliphatic aldehydes from Olea europaea . *FEMS Microbiology Letters* 198, 9-13.

Cajka, T., Riddellova, K., Klimankova, E., Cerna, M., Pudil, F., and Hajslova, J. 2010. Traceability of olive oil based on volatiles pattern and multivariate analysis. *Food Chem.* 121, 282–289.

Calín-Sánchez, A., Martínez, J. J., Vázquez-Araú, L., Burló, F., Melgarejo, P., and Carbonell-Barrachina, A. A. 2011. Volatile composition and sensory quality of Spanish pomegranates (*Punica granatum* L.). *J. Sci. Food Agric.* 91, 586–992.

European Union Commission. 1991. Regulation EEC/2568/91 on the characteristics of olive oil and olive residue oil and on the relevant methods of analysis. Official Journal of European Community, L248, 1–83.

Fujisaki, M., Endo, Y., and Fujimoto, K. 2002. Retardation of Volatile Aldehyde Formation in the Exhaust of Frying Oil by Heating Under Low Oxygen Atmospheres. *J. Am. Oil Chem. Soc.* 79, 909-914.

Galliard, T., and Matthew, J. A. 1976. The enzymic formation of long chain aldehydes and alcohols by α-oxidation of fatty acids in extracts of cucumber fruit (Cucumis sativus). *Biochim. Biophys. Acta* 424, 26–35.

García-González, D. L., Tena, N., and Aparicio, R. 2009. Contributing to interpret sensory attributes qualifying Iberian hams from the volatile profile. *Grasas Aceites* 60, 277-283.

Grimm, C. C., Lloyd, S. W., Batista, R., and Zimba, P. V. 2000. Using microwave distillation-solid-phase microextraction-gas chromatography-mass spectrometry for analyzing tissue. *J. Chromatogr. Sci.* 38, 289-296.

Grompone, M. A., Pagano, T., Pinchak, Y., and Harispe, R. 2004. Comparative study of oil deterioration during the storage of hake liver and its oil. *Grasas Aceites* 55, 291-295.

Guillén, M. D., and Uriarte, P. S. 2012. Aldehydes contained in edible oils of a very different nature after prolonged heating at frying temperature: Presence of toxic oxygenated α, β, unsaturated aldehydes. *Food Chem.* 131, 915–926.

Guinda, A., Lanzón, A., and Albi, A. 1996. Differences in hydrocarbons of virgin olive oils obtained from several olive varieties. *J. Agric. Food Chem.* 44, 1723-1726.

Hansjakob, A., Riederer, M., and Hildebrandt, U. 2011. Wax matters: absence of very-long-chain aldehydes from the leaf cuticular wax of the glossy 11 mutant of maize compromisos the prepenetration processes of Brumeria graminis. *Plant Pathology* 60, 1151-1161.

Hwang, K. T., Cuppett, S. L., Weller, C. L., and Hanna, M. A. 2002. HPLC of grain sorghum wax classes highlighting separation of aldehydes from wax esters and steryl esters. *J. Sep. Sci.* 25, 619-623.

Iyer, S., Millar, T., Clems, S., Zachgo, S., Giblin, M., and Taylor, D. 1998. In: J. Sánchez, E. Cerdá-Olmedo, and E. Martínez- Force (Eds.), *Advances in plant lipid research* (pp. 87). Sevilla: Publicaciones U. de Sevilla.

Jansen, B., Nierop, K. G. J., Hageman, J. A., Cleef, A. M., and Verstraten, J. M. 2006. The straight-chain lipid biomarker composition of plant species responsible for the dominant biomass production along two altitudinal transects in the Ecuadorian Andes. *Organic Geochem.* 37, 1514–1536.

Jurado, A., Carrapiso, A. I., Ventanas, J., and García, C. 2009. Changes in SPME-extracted volatile compounds from iberian ham during ripening. *Grasas Aceites* 60, 262-270.

Kajiwara, T., Hatanaka, A., Kawai, T., Ishihara, M., and Tsuneya, T. 1987. Long chain aldehydes in the green marine algae Ulvaceae. *Bull. Japan. Soc. Sci. Fish.* 53, 1901.

Kalua, C. M., Allen, M. S., Bedgood, D. R., Bishop, A. G., Prenzler, P. D., and Robards, K. 2007. Olive oil volatile compounds, flavour development and quality: A critical review. *Food Chem.* 100, 273–286.

Kanya, T. C. S., Raob, L. J., and Sastrya, M. C. S. 2007. Characterization of wax-esters, free fatty alcohols and free fatty acids of crude wax from sunflower seed oil refineries. *Food Chem.* 101, 1552–1557.

Kanavouras, A., Kiritsakis, A. and Hernandez, R. J. 2005. Comparative study on volatile analysis of extra virgin olive oil by dynamic headspace and solid phase micro-extraction. *Food Chem.* 90, 69–79.

Koprivnjak, O., Moret, S., Populin, T., Lagazio, C., and Conte, L. S. 2005. Variety differentiation of virgin olive oil based on n-alkane profile. *Food Chem.* 90, 603–608.

Kubo, A., Lunde, Ch. S., and Kubo, I. 1995. Antimicrobial activity of the olive oil flavor compounds. *J. Agric. Food Chem.* 43, 1629-1633.

López-Vázquez, C., Orriols, I., Perelló, M. C., and de Revel, G. 2012. Determination of aldehydes as pentafluorobenzyl derivatives in grape pomace distillates by HS-SPME-GC/MS. *Food Chem.* 130, 1127-1133.

Margenat, L., Jachmanián, I., and Grompone, M. A. 2005. Determination of lipoxygenase activity in hake liver. *Grasas Aceites.* 56, 205-208.

McDonald, R. E., Nordby, H. E., and McCollum, T. G. 1993. Epicuticular Wax Morphology and Composition are Related to Grapefruit Chilling Injury *Hortscience* 28, 311-312.

Martín, L., Timón, M. L., Petrón, M. J., Ventanas, J., and Antequera, T. 2000. Evolution of volatile aldehydes in Iberian ham matured under different processing conditions. *Meat Sci.* 54, 333-337

Martin, R. O., and Stumpf, P. K. 1959. Fat metabolism in higher plants. XII. α-Oxidation of long-chain fatty acids. *J. Biol. Chem.* 234, 2548-2554.

Mayuoni-kirshinbaum, L., and Tietel, Z. 2012. Identification of aroma-active compounds in 'wonderful' pomegranate fruit using solvent-assisted flavour evaporation and headspace solid-phase micro-extraction methods *Eur. Food Res. Technol.* 235, 277–283.

Melgarejo, P., Sánchez, A. C., Vázquez-Araújo, L., Hernández, F., Martínez, J., Legua, P., and Carbonell-Barrachina, A. A. 2011. Volatile composition of pomegranates from 9 Spanish cultivars using headspace solid phase microextraction. *J. Food Sci.* 76, S114–S120.

Morris, B. D., Foster, S. P., and Harris, M. O. 2000. Identification of 1-octacosanal and 6-methoxy-2-benzoxazolinone from wheat as ovipositional stimulants for Hessian fly, Mayetiola destructor. *J. Chem. Ecol.* 26, 859–873.

Munch, J. W., Munch, D. J., Winslow, S. D., Wendelken, S. C., and Pepich, B. V. 1998. Epa Method 556: Determination of Carbonyl Compounds in Drinking Water by Pentafluorobenzylhydroxylamine Derivatization and Capillary Gas Chromatography with Electron Capture Detection, US Environmental Protection Agency, Cincinnati, OH.

Narváez, M., Gallardo, E., Ríos, J. J., and León-Camacho, M. 2010. A tentative characterization of volatile compounds from iberian dried-cured ham according to different anatomical locations. A detailed study. *Grasas Aceites* 61, 369-377.

Nass, R., Markstadter, C., Hauke, V., Riederer, M. 1998. Quantitative gas chromatographic analysis of plant cuticular waxes containing long-chain aldehydes. *Phytochem. Analysis.* 9, 112-118.

Nordby, H. E., and R. E. McDonald. 1990. Squalene in grapefruit wax as a possible natural protectant against chilling injury. *Lipids* 25, 807-810.

Nordby, H. E., and McDonald, R. E., 1991. Relationship of Epicuticular Wax Composition of Grapefruit to Chilling Injury. *J. Agric. Food Chem.* 39, 957-962.

Pérez-Camino, M. C., Moreda, W., Mateos, R., and Cert, A. 2003. Simultaneous determination of long-chain aliphatic aldehydes and waxes in olive oils. *Journal Chromatography* A 983, 283–288.

Pérez-Camino, M. C., Gómez-Coca, R. B., and Moreda, W. 2012. Waxy fraction containing long-chain aliphatic aldehydes in virgin olive oils. *Food Chem.* 132, 1451–1456.

Piccirillo, P., Fasano, P., Mita, G., De Paolis, A., and Santino, A. 2005. Exploring the role of lipoxygenases on walnut quality and shelf-life. *Acta Hort. (ISHS)* 705, 543-545.

Reina-Pinto, J. J., and Yephremov, A. 2009. Surface lipids and plant defenses. *Plant Physiol. Biochem.* 47, 540–549.

Riendeau, D., and Meighen, E. 1985. Enzymatic reduction of fatty acids and acyl-CoAs to long chain aldehydes and alcohols. *Experientia* 41, 707–713.

Reisige, K., Gorzelanny, C., Daniels, U., and Moerschbacher, B. M. 2006. The C28 aldehyde octacosanal is a morphogenetically active component involved in host plant recognition and infection structure differentiation in the wheat stem rust fungus. *Physiol. Mol. Plant Pathol.* 68, 33-40.

Stephanou, E. 1989. Long-Chain n-Aldehydes An Overlooked but Ubiquitous Compound Class of Possible Geochemical and Environmental Significance. *Naturwissenschaften* 76, 464- 467.

Thomae, C. 1913. Plant Fats *Journal für praktische Chemie* 87, 144.

Tijet, N., Schneider, C., Muller, B. L., and Brash, A. R. 2001. Biogenesis of volatile aldehydes from fatty acid hydroperoxides: molecular cloning of a hydroperoxide lyase (CYP74C) with specificity for both the 9- and 13-hydroperoxides of linoleic and linolenic acids. *Arch. Biochem. Biophys.* 386, 281-289.

Varlet, V., Prost, C. and Serot, T. 2007. Volatile aldehydes in smoked fish: Analysis methods, occurrence and mechanisms of formation. *Food Chem.* 105, 1536–1556.

Vichi, S., Castellote, A. I., Pizzale, L., Conte, L. S., Buxaderas, S., and López-Tamames, E. 2003. Analysis of virgin olive oil volatile compounds by headspace solid-phase microextraction coupled to gas chromatography with mass spectrometric and flame ionization detection. *J. Chromatogr.* A 983, 19-33.

Warner, K., Evans, C. D., List, G., Dupuy, H. P., Wadsworth, J. I., and Goheen, G. E. 1978. Flavor Score correlation with pentanal and hexanal contents of vegetable oil. *J. Am. Oil Chem. Soc.* 55, 252-256.

Wood, R., and Harlow, R. D. 1969. Gas-liquid chromatographic analysis of free long-chain aldehydes. *J. Lipid Research* 10, 463-465.

Yasuhara, A., and Shibamoto, T. 1995. Quantitative Analysis of Volatile Aldehydes Formed from Various Kinds of Fish Flesh during Heat Treatment. *J. Agric. Food Chem.* 43, 94–97.

Yokouchi, Y., Mukai, H., Nakajima, K., and Ambe, Y. 1990. Atmospheric Environment. Part A. *General Topics* 24, 439–442.

In: New Developments in Aldehydes Research
Editors: L. Torrioni and E. Pescasseroli
ISBN: 978-1-62417-090-4

Chapter 4

INHIBITION OF MICROBIAL BIOCATALYSTS BY BIOMASS-DERIVED ALDEHYDES AND METHODS FOR ENGINEERING TOLERANCE

Laura R. Jarboe[1*] and Zhanyou Chi[2#]
[1]Department of Chemical and Biological Engineering,
Iowa State University, Iowa, US
[2]Center for Sustainable Environmental Technologies,
Iowa State University, Iowa, US

ABSTRACT

In our effort to fermentatively produce biorenewable fuels and chemicals that are economically competitive with petroleum, it is desirable to use sugars derived from lignocellulosic biomass. However, the recalcitrance of this biomass requires some sort of depolymerization treatment to release the fermentable substrates. This depolymerization can be performed by enzymes, ionic liquids, hydrolysis or pyrolysis. Each depolymerization method has its own benefits and challenges. The challenge that is described here is the fact that hydrolysis and pyrolysis both result in the production of compounds that are inhibitory to the biocatalyst, limiting utilization of these carbon- and energy-rich streams. Specifically, we consider biocatalyst inhibition by aldehydes such as furfural, 5-hydroxymethylfurfural (5-HMF), glycolaldehyde, syringaldehyde, vanillin and hydroxybenzaldehyde. Technologies exist for selectively removing these inhibitors, such as overliming or treatment with laccase enzymes, but these potentially increase the process cost. Selection-based strain improvement of a variety of industrially-relevant biocatalysts, including yeast and *Escherichia coli*, has provided insight into methods for strain improvement. Transcriptomic studies have provided complementary insights. Here, we describe these efforts and summarize both what is known regarding the mechanism of growth inhibition by these aldehydes and how biocatalysts can be engineered for increased tolerance.

* Corresponding author: Laura R. Jarboe, Department of Chemical and Biological Engineering, Iowa State University, ljarboe@iastate.edu.
# Zhanyou Chi, Center for Sustainable Environmental Technologies, Iowa State University, chiz@iastate.edu.

INTRODUCTION

There are a variety of motivations for developing technologies for the production of fuels and chemicals from biomass. These motivations include, but are not limited to, increased energy and food security, sustainability, climate change and costs [1-5]. One general method for the production for these biorenewable chemicals is to release energy-rich sugars from biomass and use microbial biocatalysts to ferment these sugars to biorenewable fuels and chemicals. These microbial biocatalysts are often subjected to metabolic engineering to improve strain characteristics [6, 7]. Glucose is the sugar preferred by most microbes and it can be easily released from sugar-rich crops, such as sugar cane, beets and corn [8]. The fermentative production of a wide variety of fuels and chemicals from glucose at high yields, titers and productivities has been reported [6, 7]. However, a variety of concerns, including food security and land use, make lignocellulosic biomass an attractive source of carbon and energy for these microbes instead of dedicated sugar crops.

While the vast abundance of lignocellulosic biomass, as well as its ability to grow in locations that are generally non-arable, make it a very attractive source of carbon and energy, it is generally in a form that cannot be readily used by microbes. This means that some sort of pre-processing is required in order to make the energy-rich sugars accessible for fermentation. These pretreatment methods include digestion with enzymes or ionic liquids, hydrolysis and pyrolysis [9-13]. Each of these methods has their advantages and disadvantages.

Sugar-rich streams produced by hydrolysis and pyrolysis are both known to contain compounds that are inhibitory to microbial biocatalysts [12, 14]. This means that either additional process cost must be incurred to remove the inhibitory compound(s) or the microbe must be modified for increased tolerance of these compounds. Weak organic acids, such as acetate, and aldehydes, such as furfural and 5-hydroxymethylfurfural (5-HMF), have been intensively studied as model inhibitors [15-17]. Here we review the research related to these aldehydes, including their production during biomass processing, methods for their removal and methods to increase biocatalyst tolerance to their inhibitory properties.

FORMATION AND RELEASE OF ALDEHYDES DURING BIOMASS PROCESSING

A variety of aldehydes are produced during both hydrolytic and pyrolytic biomass processing. Select representative data are shown in Tables 1 and 2 and chemical structures are given in Table 3. These aldehydes can be chemically produced during the biomass processing or are already present in the lignin portion of the biomass. For example, xylose dehydration is known to produce furfural, a process which is used for industrial furfural production [18]. Other aldehydes, particularly the aromatics syringaldehdye and vanillin, can be directly derived from lignin [19-21]. The identity and amount of aldehydes produced can vary according to many variables, including biomass identity and processing condition(s). As shown in Table 1, dilute acid hydrolysis of spruce produces roughly two-fold less furfural than dilute acid hydrolysis of alder at the same solids loading value [22].

Patwardhan *et al.* [23] showed a relationship between the concentration of aldehydes in pyrolyzed pure cellulose and pre-treatment with metal chlorides. Specifically, the addition of

0.2 mmole $MgCl_2$ per g of cellulose increased the furfural concentration to 4 wt%, relative to 1 wt% for the no-$MgCl_2$ control. Contrastingly, there was only a slight increase in 5-HMF content and a slight decrease in glycolaldehyde. These results also highlight the *de novo* formation of these aldehydes from pure cellulose. Thus, aldehyde formation during biomass processing is complex and difficult to predict or minimize.

Table 1. Aldehydes detected in biomass hydrolysate

Biomass	Aldehyde	Aldehyde Concentration (g/L)	Initial % solids	Source
Dilute acid hydrolysis				
alder	furfural	2.7 - 3.3	33	[22]
aspen		2.1 - 3.5	33	[22]
birch		1.6 - 3.8	25	[22]
		0.5 - 4.6	33	[22]
willow		0.4 - 1.8	25	[22]
		0.3 - 3.2	33	[22]
pine		0.1 - 1.3	25	[22]
		0.7 - 1.7	33	[22]
spruce		0.2 - 1.0	25	[22]
		0.4 - 1.4	33	[22]
alder	5-HMF	2.6 - 4.5	33	[22]
aspen		1.3 - 6.8	33	[22]
birch		0.8 - 4.3	25	[22]
		0.2 - 5.8	33	[22]
willow		0.3 - 3.9	25	[22]
		0.6 - 3.8	33	[22]
pine		1.0 - 6.9	25	[22]
		1.7 - 8.6	33	[22]
spruce		1.5 - 6.7	25	[22]
		1.5 - 8.4	33	[22]
corn stover	furfural	1.8	10	[59]
	5-HMF	0.09	10	[59]
	syringaldehyde	0.01	10	[59]
	vanillin	0.01	10	[59]
	hydroxybenzaldehyde	0.01	10	[59]
	dihydroxybenzaldehyde	0.01	10	[59]
	5-HMF	4.0	20	[60]
	furfural	3.9	20	[60]
Alkaline wet oxidation				
wheat straw	furfural	0.0 - 0.088	6	[61]
	5-HMF	0.0 - 0.01	6	[61]
	syringaldehyde	0.001 - 0.045	6	[61]
	vanillin	0.005 - 0.058	6	[61]
	hydroxybenzaldehyde	0.007 - 0.035	6	[61]
Dilute acid pretreatment and enzymatic hydrolysis				
bagasse	furfural	0.6 - 1.0	16	[62]
	5-HMF	1.6 - 3.0	16	[62]
spruce	furfural	0.7 - 1.2	16	[62]
	5-HMF	1.7 - 3.2	16	[62]
bagasse	furfural	1.1 - 1.6	*n/a*	[63]
	5-HMF	0.2 - 0.5	*n/a*	[63]
pine	furfural	0.5	10	[64]
	5-HMF	0.9	10	[64]

Table 2. Aldehydes detected in biomass-derived bio-oil. Due to the non-aqueous nature of the pyrolysis product, concentrations are given as a mass percentage instead of g/L

Biomass	Aldehyde	Aldehyde Concentration (%)	Source
cellulose	furfural	0.3 – 1.3	[65]
	glycolaldehyde	5.4 – 6.7	[65]
	5-methyl-2-furfural	0.1 - 0.2	[65]
	5-HMF	0.0 - 2.8	[65]
beech wood	furfural	1.58 - 2.52	[66]
	methyl-furfural	0.36 - 0.65	[66]
	methyl-2-furancarboxaldehyde	0.38 - 0.74	[66]
spruce wood	furfural	1.63 - 2.38	[66]
	methyl-furfural	0.33 - 0.84	[66]
	methyl-2-furancarboxaldehyde	0.41 - 0.71	[66]
olive husk	furfural	0.61 - 1.76	[66]
	5-methyl-2-furfural	0.26 - 0.71	[66]
	methyl-2-furancarboxaldehyde	0.24 - 0.48	[66]
hazelnut shell	furfural	0.60 - 1.86	[66]
	5-methyl-2-furfural	0.30 - 0.75	[66]
	methyl-2-furancarboxaldehyde	0.26 - 0.50	[66]
bamboo	furfural	1.72 - 1.74	[67]
	methylbenzaldehyde	0.31 - 8.02	[67]
	hydroxybenzaldehyde	1.05 - 0.94	[67]
rice straw	furfural	0.62 - 6.66	[67]
	methylbenzaldehyde	4.17	[67]
	vanillin	0.65 - 0.99	[67]
	hydroxybenzaldehyde	0.28 - 0.42	[67]
poplar	furfural	0.52	[68]
	syringaldehyde	0.50	[68]
	vanillin	0.3	[68]
	glycolaldehyde	1.62	[68]
Pterocarpus indicus (New Guinea rosewood)	furfural	9.06	[69]
	5-methyl-2-furfural	1.82	[69]
	vanillin	6.35	[69]
	salicylaldehyde	2.70	[69]
	syringaldehyde	4.54	[69]
	cinnamaldehyde	2.19	[69]
forest residue	furfural	0.24	[70]
	glycolaldehyde	6.45	[70]

Table 3. Chemical structure of aldehydes detected in biomass hydrolysate and pyrolysate. Structures were obtained from ChemID Plus

Aldehyde	Structure	Molecular Weight	CAS Number
furfural	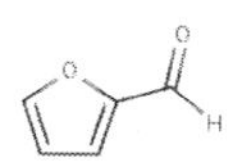	96.08	98-01-1
5-hydroxymethylfurfural (5-HMF)	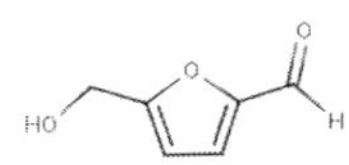	126.11	67-47-0
5-methyl-2-furfural	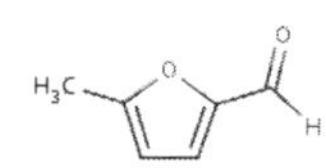	110.11	620-02-0
glycolaldehyde	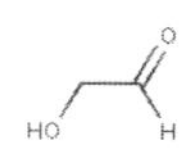	60.05	141-46-8
syringaldehyde	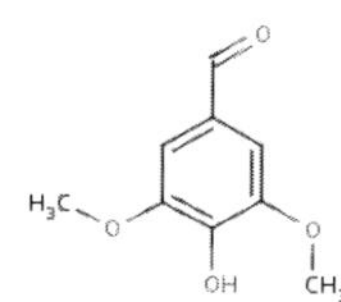	182.17	134-96-3
vanillin	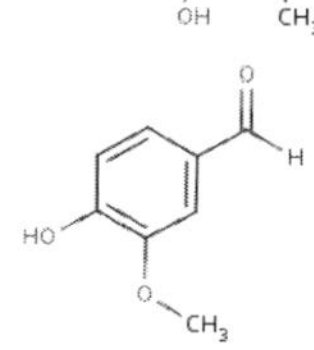	152.15	121-33-5
salicylaldehyde	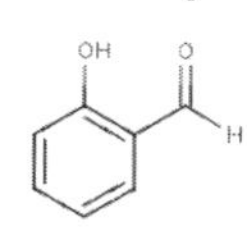	122.12	90-02-8
4-hydroxybenzaldehyde	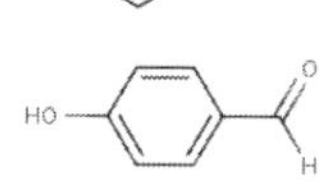	122.12	123-08-0
cinnamaldehyde	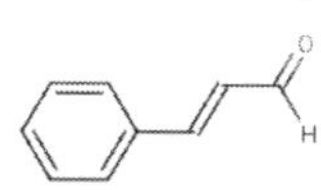	132.16	104-55-2
3,4-dihydroxybenzaldehyde	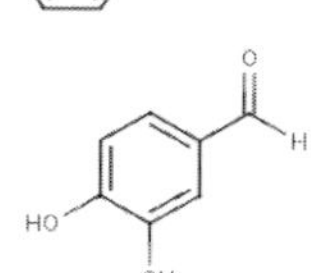	138.12	139-85-5

INHIBITION OF MICROBIAL BIOCATALYSTS BY ALDEHYDES

The primary motivation for biomass hydrolysis and pyrolysis is to release and concentrate energy-rich compounds that can serve as the substrate for production of biorenewable fuels and chemicals. However, fermentative utilization of these energy-rich compounds by microbial biocatalysts is stymied by the aldehyde inhibitors. Data for select inhibitors and ethanologenic microbial biocatalysts are given in Tables 4 and 5. Other excellent studies include, but are not limited to [24-26].

As shown in Tables 1 and 2, the furfural concentration in biomass hydrolysate or pyrolysate is frequently in excess of 1 g/L, a concentration that is sufficient to inhibit the growth and product formation of many ethanologenic microbes by more than 25% (Table 4). Similarly, the abundance of vanillin in the pyrolysis product (Table 2) is sufficiently high to anticipate biocatalyst inhibition.

Microbial sensitivity to these aldehydes varies greatly according to microbe species, aldehyde identity and growth condition. For example, the yeast species *C. shehatae* is relatively robust to furfural when compared to the yeast *S. cerevisiae*: approximately 60% of *C. shehatae* growth and ethanol production remain in the presence of 10 mM furfural (1 g/L) while only 20% remains in *S. cerevisiae* (Table 4). Note however that both species suffer an approximately 90% loss in growth and metabolism in the presence of 20 mM furfural (2 g/L). *P. stipitis* shows a greater than 90% loss in growth in the presence of approximately 10 mM 4-hydroxybenzaldehyde, vanillin or syringaldehyde but is not nearly so sensitive to furfural or 5-HMF (Tables 4 and 5). This difference is consistent with Zaldivar *et al.*'s observation that aromatic aldehydes were at least twice as toxic to *E. coli* as furfural or 5-HMF, where evaluation was performed on a weight basis [16]. Miller *et al.* showed differences in *E. coli* growth sensitivity in response to growth media additives, as described below [27].

These varying sensitivities can probably be attributed to differences in both the mechanism of the inhibition and the inherent microbial defense systems. Zaldivar *et al.*'s *E. coli* studies led to the conclusion that toxicity was directly related to the hydrophobicity of the focal aldehyde, though there was no evidence of membrane damage [16].

Table 4. Inhibition of biocatalyst growth and metabolism by furfural, 5-HMF and glycolaldehyde

Aldehyde	Concentration		Biocatalyst	Relative cell growth (%)	Relative ethanol production (%)	Source
	g/L	mM				
furfural	0.5	5	*Candida shehatae*	81	80	[71]
	1	10		62	53	[71]
	2	20		9.7	9.6	[71]
	0.5	5	*Pichia stipitis*	75	71	[71]
	1	10		53	29	[71]
	2	20		1	5	[71]
	0.5	5	*Saccharomyces cerevisiae*	53	57	[71]
	1	10		19	20	[71]
	2	20		10	11	[71]
	0.5	5	*Zymomonas mobilis*	82	96	[71]
	1	10		81	82	[71]
	2	20		44	56	[71]
	3	30	*Kluyveromyces marxianus*	50	50	[72]
	2.0	21	*Escherichia coli*	75	*n/a*	[16]
	2.4	25		50	*n/a*	[16]
	3.5	36		0	*n/a*	[16]
	5.0	52	*Corynebacterium glutamicum*	*n/a*	82	[73]

Aldehyde	Concentration		Biocatalyst	Relative cell growth (%)	Relative ethanol production (%)	Source
	g/L	mM				
5-HMF	1	8	*C. shehatae*	92	67	[71]
	3	20		32	10	[71]
	5	40		8	4	[71]
	1	8	*P. stipitis*	95	83	[71]
	3	20		31	10	[71]
	5	40		1.4	8.6	[71]
	1	8	*S. cerevisiae*	35	29	[71]
	3	20		17	17	[71]
	5	40		11	5	[71]
	1	8	*Z. mobilis*	51	85	[71]
	3	20		69	87	[71]
	5	40		33	47	[71]
	4	30	*K. marxianus*	50	50	[72]
	2.3	18	*E. coli*	75	*n/a*	[16]
	2.7	21		50	*n/a*	[16]
	4.0	32		0	*n/a*	[16]
	10.0	79.3	*C. glutamicum*	*n/a*	62	[73]
glycol-aldehyde	0.03	0.5	*S. cerevisiae*	91	n/a	[74]
	0.12	2.0		n/a	50	[74]
	0.30	5.0		43	n/a	[74]
	1.2	20		4	n/a	[74]

Table 5. Inhibition of biocatalyst growth and metabolism by the aromatic aldehydes 4-hydroxybenzaldehyde, vanillin and syringaldehyde

Aldehyde	Concentration		Biocatalyst	Relative cell growth (%)	Relative ethanol production (%)	Source
	g/L	mM				
4-hydroxy-benzaldehyde	0.5	4.1	*P. stipitis*	57	30	[71]
	0.75	6.1		30	16	[71]
	1.5	12		0	4	[71]
	0.5	4.1	*S. cerevisiae*	75	97	[71]
	0.75	6.1		47	63	[71]
	1.5	12		13	25	[71]
	0.5	4.1	*Z. mobilis*	16	21	[71]
	0.75	6.1		8	14	[71]
	1.5	12		8	11	[71]
	0.15	1.2	*E. coli*	75	*n/a*	[16]
	0.30	2.5		50	*n/a*	[16]
	1.25	10.2		0	*n/a*	[16]
vanillin	0.5	3.3	*P. stipitis*	12	11	[71]
	1	6.6		0.7	9	[71]
	2	13		1.4	6	[71]
	0.5	3.3	*S. cerevisiae*	49	70	[71]
	1	6.6		14	17	[71]
	2	13		9	11	[71]
	0.5	3.3	*Z. mobilis*	62	86	[71]
	1	6.6		37	74	[71]
	2	13		12	20	[71]
	0.4	2.6	*E. coli*	75	*n/a*	[16]
	0.5	3.3		50	*n/a*	[16]
	1.5	9.9		0	*n/a*	[16]
syring-aldehyde	0.2	1.1	*P. stipitis*	72	61	[71]
	0.75	4.1		38	20	[71]

Table 5. (Continued)

Aldehyde	Concentration		Biocatalyst	Relative cell growth (%)	Relative ethanol production (%)	Source
	g/L	mM				
	1.5	8.2		3.6	6	[71]
	0.2	1.1	*S. cerevisiae*	100	74	[71]
	0.75	4.1		39	46	[71]
	1.5	8.2		19	33	[71]
	0.2	1.1	*Z. mobilis*	82	97	[71]
	0.75	4.1		72	101	[71]
	1.5	8.2		60	83	[71]
	0.30	1.6	*E. coli*	75	*n/a*	[16]
	0.60	3.3		50	*n/a*	[16]
	2.5	14		0	*n/a*	[16]

Zaldivar's *E. coli* research also investigated the effect of binary combinations of aldehydes and found that toxicity was at least additive, with combinations including furfural being more toxic than the expected additive combination [16].

Thus, many of the aldehydes which have been quantified in biomass breakdown products and have been tested with a variety of biocatalysts are abundant enough to cause concern regarding biocatalyst inhibition. Given its ubiquity and ability to enhance the toxicity of other aldehydes, furfural has been considered by the scientific community as a model inhibitory aldehyde and is described in more detail below.

SELECTIVE REMOVAL OF ALDEHYDES

A variety of detoxification methods have been developed to remove contaminant compounds in biomass hydrolysate, as described in several reviews [28-30]. These treatment methods can be roughly classified into chemical, biological, and physical treatment. Some representative treatment methods and their effectiveness for aldehyde removal are listed in Table 6.

As an example of a chemical treatment, the addition of calcium hydroxide, *i.e.* overliming, removes inhibitory compounds from biomass hydrolysate via precipitation. In one report, overliming was able to remove up to 85% of furfural and up to 60% of dihydroxbenzaldehydes [31, 32] (Table 6). Similarly, ammonium hydroxide addition removed up to 50% of furfural and up to 89% of benzaldehydes, such as vanillin [32]. In both cases, the amount of calcium hydroxide or ammonium hydroxide added, and thus the final pH attained during treatment, has a significant impact on the effectiveness (and cost) of the treatment. Note that overly aggressive overliming can also result in the loss of sugars [33].

Laccase enzymes, also known as "blue copper oxidase" have historically been used in the food and beverage industry to degrade polyphenols; more than 2,200 laccase enzymes were classified in the construction of the Laccase Engineering Database [34]. These enzymes have also shown promise as a biological treatment to degrade the inhibitory aldehydes present in biomass hydrolysate [35]. Such treatments have been shown to remove more than 90% of benzaldehydes [36], though there is a lack of recorded activity against furfural and 5-HMF. We propose that the effectiveness of laccase treatments could possibly be improved through selection of the appropriate enzymes or modification of enzyme structure and activity.

Table 6. Detoxification methods for aldehyde removal from biomass hydroylsate

Treatment method	Aldehyde	% removed	Source
Calcium hydroxide (overliming)	furfural	31 - 32	[32]
		57	[33]
		85	[31]
		52 - 70	[75]
		20	[37]
	5-HMF	29 - 31	[32]
		55	[33]
		22	[37]
		43 - 66	[75]
	4-hydroxybenzaldehyde	39 - 41	[32]
	vanillin	3	[32]
	3,4-dihydroxybenzaldehyde	53 - 60	[32]
	trans-cinnamaldehyde	50	[32]
Ammonium hydroxide	furfural	42 - 50	[62]
		38 - 45	[32]
	5-HMF	47 - 50	[62]
		38 - 44	[32]
	4-hydroxybenzaldehyde	13 - 17	[32]
	vanillin	88 - 89	[32]
	3,4-dihydroxybenzaldehyde	38 - 45	[32]
	trans-cinnamaldehyde	0	[32]
Sodium sulfite	furfural	0 - 53	[37]
		16	[62]
	5-HMF	0 - 52	[37]
		0 - 9	[62]
Laccase	furfural	0	[37]
	5-HMF	0	[37]
	vanillin	98	[36]
	4-hydroxybenzaldehyde	>90	[36]
Evaporation	furfural	47 - 100	[37]
		>58	[76]
	5-HMF	0 - 4	[37]
	vanillin	29	[38]
Extraction with ether	furfural	100	[38]
	vanillin	100	[38]
Polymeric adsorbent	furfural	93.3 - 94.5	[77]
	furfural	23.1 - 88.3	[64]
	5-HMF	13.4 - 66.4	[64]
Anion exchange	furfural	31 - 73	[37]
	5-HMF	26 - 70	[37]

In terms of physical removal of inhibitory aldehydes, vacuum evaporation can effectively remove volatile aldehydes such as furfural, but its effectiveness against non-volatile aldehydes, such as 5-HMF, is low [37]. Extraction of hydrolysate or pyrolysate with organic solvents has proven effective. For example, treatment of hydrolysate with ether was able to remove most of the furfural and vanillin [38].

Adsorption on anion exchange resin is an effective method to remove furfural and 5-HMF, but there are concerns about the high cost [30]. Adsorption with polymers such as polyelectrolyte polyethyleneimine (PEI) was able to remove 88.3% of the furfural and 66.4%

of the 5-HMF. Also, furfural and 5-HMF can be recovered after removal by washing with dilute sulfuric acid solution. This is an emerging approach to recover biochemcials from biomass hydrolysate [39].

There is less information currently available regarding removal of aldehydes from biomass pyrolysate, but extraction of bio-oil with a variety of solvents, such as tri-n-octylamine in conjunction with 1-octanol, improved bio-oil fermentability by yeast, though it is not clear if this was due to removal of aldehydes [40]. Similarly, overliming improved bio-oil fermentability, though the extent of aldehyde removal is unknown at this time [40].

Thus, a variety of pretreatment methods exist to selectively remove aldehydes and other inhibitors from biomass hydrolysate and pyrolysate. These treatments can be selected and fine-tuned based on the needs of the process. However, each treatment step added to the process results in an increase in the process cost, an increase that may not be justified by the resulting increase in fermentability of the sugar-rich stream. Another alternative is to increase the robustness of the biocatalyst, so that the need for aldehyde removal is reduced or eliminated entirely.

Increasing Microbial Tolerance to Aldehydes

While the technologies described above enable the selective removal of inhibitory contaminant compounds, another method is to increase the robustness of the microbe to the relevant inhibitors. This need for improved biocatalyst robustness is pervasive within the entire framework for fermentative production of biorenewable fuels and chemicals [41]. Most generally speaking, there are two methods for increasing biocatalyst robustness. The first involves an evolutionary approach, where populations are enriched for either naturally-occurring mutations or expression of gene libraries under selective pressure of the focal inhibitor. The second approach involves rational, predictive genetic modification to increase the tolerance of the biocatalyst to the inhibitory compound; this approach requires a basic understanding of why the focal compound is inhibitory. Here we describe both approaches in a variety of industrially-relevant biocatalysts.

Ethanologenic *E. coli*

E. coli W had been previously engineered for ethanol production via expression of the *Z. mobilis* pyruvate to ethanol (PET) pathway and then evolved for high ethanol production/sugar consumption in defined mineral salts media [42, 43]. A derivative of this strain, designated LY180, was selected as a model strain for understanding furfural tolerance by both identification of the mechanism of inhibition and evolution and reverse engineering of an evolved strain. For this strain in a growth condition of defined mineral salts media with 10% (w/v) xylose, addition of 1.0 g/L furfural completely inhibited ethanol production [44]. Previous analysis had demonstrated that *E. coli* possesses an NADPH-dependent furfural reductase that converts furfural to the less-inhibitory furfuryl alcohol [45], but the identity of the gene(s) encoding this activity remained unknown. With the goal of identifying the pathways perturbed by furfural challenge, transcriptome analysis was performed immediately before and 15 minutes after addition of 0.5 g/L furfural [27]. This analysis showed 2-fold or greater perturbation of approximately 400 genes; network component analysis of the transcriptome data identified 21 transcription factors as having significantly altered regulatory activity. The fact that the O-acetyl-L-serine-responsive regulator CysB had increased activity while the S-adenosyl-methionine-responsive regulator MetJ had decreased activity suggested an accumulation of the former metabolite and depletion of the latter. As part of the cysteine and methionine biosynthesis pathway, O-acetyl-L-serine is a metabolite upstream of S-adenosyl-methionine (Figure 1). Entry of O-acetyl-L-serine into this pathway requires hydrogen sulfide, in the reaction that produces cysteine. Production of hydrogen sulfide requires 3 NADPH; this is a very high demand for a precious cellular resource. Given that furfural reduction was previously observed to be NADPH-dependent, Miller *et al.* proposed that depletion of the NADPH pool via furfural reduction limited the production of hydrogen sulfide, causing the observed accumulation of O-acetyl-L-serine and depletion of S-adenosyl-methionine.

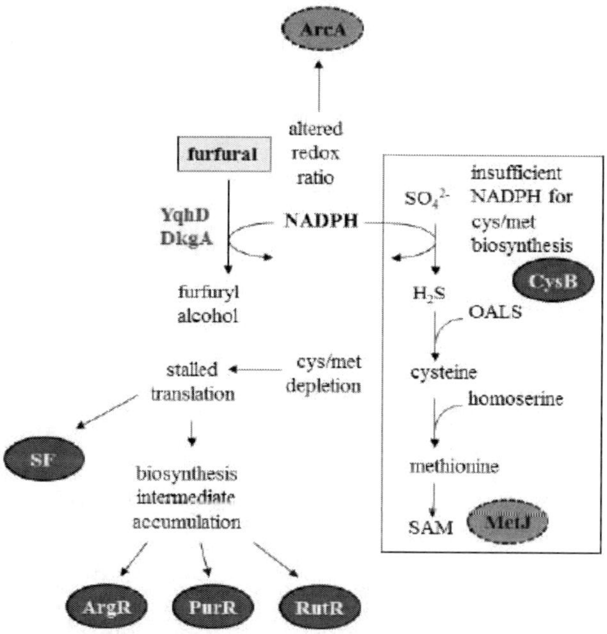

Figure 1. The metabolic and transcriptomic response of ethanologenic *E. coli* strain LY180 to challenge with furfural. Furfural reduction by YqhD and DkgA depletes NADPH levels, limiting production of H_2S and thereby limiting biosynthesis of the cysteine and methionine needed for protein translation and growth. Ovals indicate transcription factors, where red with a solid border indicates an increase in regulatory activity, green with a dashed border indicates decreased regulatory activity. OALS, O-acetyl-L-serine; SAM, S-adenosyl-methionine. Furfural tolerance can be increased by supplementing with compounds that mitigate the need for H_2S, by increasing NADPH availability through use of glucose or expression of a transhydrogenase enzyme or by silencing of the furfural reductase enzymes.

This decrease in hydrogen sulfide production would also limit production of the amino acids cysteine and methionine, limiting protein translation and thereby limiting growth. This hypothesis was supported by the fact that supplementation with D-cysteine, which can be used to produce hydrogen sulfide by an NADPH-independent reaction, increased furfural tolerance [27]. Similarly, the use of glucose as carbon source instead of xylose also increased furfural tolerance, where glucose utilization enables production of NADPH through the oxidative branch of the pentose phosphate pathway [27]. This finding that increasing the availability of reduced sulfur compounds or increasing the availability of NADPH both increase furfural tolerance can be used to tailor the fermentation condition to circumvent furfural toxicity. For example, Nieves *et al.* showed that supplementation with reduced sulfur compounds, particularly sodium metabisulfite, significantly increased utilization of hydrolyzed sugarcane baggase by ethanologenic *E. coli* [46].

It is also desirable to use the identified toxicity mechanism to improve the biocatalyst for utilization of biomass-derived sugars. With this goal in mind, Miller *et al.* increased expression of the native membrane-associated transhydrogenase enzyme, encoded by *pntAB* [27]. This enzyme can convert NADH, which is relatively abundant during fermentative growth, to the relatively scarce NADPH [47]. By increasing the availability of NADPH, furfural tolerance in defined minimal xylose media was increased [27].

In a complementary approach, Miller *et al.* also performed metabolic evolution for furfural tolerance of strain LY180 in defined minimal media with 10% w/v xylose [44]. Furfural concentrations started at 0.5 g/L and were gradually increased to 1.3 g/L over the course of 53 serial transfers; the final strain was dubbed EMFR9. In the presence of 1.0 g/L furfural EMFR9 showed growth and ethanol production that were indistinguishable from the no-furfural control. EMFR9's increased furfural tolerance also resulted in increased tolerance and utilization of hydrolyzed sugarcane bagasse, suggesting that furfural toxicity is mainly responsible for the toxicity of hydrolyzed bagasse [44]. While this isolation of a furfural-tolerant mutant is useful, it is even more impactful to identify the mechanism of furfural tolerance acquired by this strain so that additional insight can be gained into furfural toxicity and similar modifications can be implemented into other biocatalysts. Therefore, both phenotypic and transcriptomic studies were used to reverse engineer this furfural-tolerant mutant. Phenotypic studies showed that the evolved strain EMFR9 had an approximately 50% decrease in the *in vivo* furfural reduction rate relative to the parent strain LY180 [44]. Without knowing the identity of the enzyme responsible for furfural reduction, it is difficult to investigate the cause of this decreased reduction rate. Therefore, transcriptome analysis focused on identification of oxidoreductase-encoding genes with decreased expression in the evolved strain relative to the parent [44]. Four candidate genes were identified: *yqhD*, *dkgA*, *yjjN* and *yqfA*. Additional studies revealed that YqhD and DkgA both have NADPH-dependent furfural reductase activity and that their deletion from the parental strain LY180 increased furfural tolerance and reinstatement of high expression in the evolved strain decreased furfural tolerance [44]. Prior to this study, it was not known that YqhD was the major *E. coli* furfural reductase; this enzyme has since proved to be important in a variety of biorenewable-related projects [48]. Further studies revealed that an insertion mutation within the regulatory gene *yqhC* was responsible for transcriptional silencing of *yqhD* [49].

Therefore, the evolutionary approach revealed that one mechanism for increasing tolerance is to stop the depletion of NADPH by decreasing the expression of NADPH-dependent furfural reductase enzymes. Thus, the transcriptomic studies of the parent strain and reverse engineering of the evolved strain both reached the same conclusion regarding furfural toxicity in *E. coli* in this growth condition: depletion of NADPH via furfural reduction is the key problem. As an alternative to supplementation with reduced sulfur compounds or conversion of NADH to NADPH, it has also been shown that the NADH-dependent furfural reductase FucO can effectively convert furfural to the less toxic furfuryl alcohol without depleting the NADPH needed for biosynthesis [50]. In this manner, expression of FucO in conjunction with silencing of *yqhD* can enable *E. coli* tolerance of up to 1.4 g/L furfural with no discernible difference from the no-furfural control condition [50].

Other studies have shown that furfural can alter DNA structure and directly cause mutations [51, 52]. While interactions with DNA did not appear to be problematic in the studies described above, enrichment studies with similar furfural concentrations showed that increased expression of the thymidylate synthase-encoding *thyA* gene conferred furfural tolerance [53]. Note that ThyA is involved in *de novo* synthesis of dTMP, a metabolite that is critical for DNA synthesis and repair [54]. Supplementation of thymine or thymidine increased furfural tolerance in a manner similar to overexpression of *thyA* [53]. Other genes that contribute to furfural tolerance include cryptic gene *ucpA*, for reasons that remain unclear [50]. When tested, methods for increasing furfural tolerance often apply to 5-HMF as well. For example, the evolved furfural-resistant EMFR9 also has increased tolerance of 5-HMF [55].

Yeast

Heer *et al.* performed directed evolution of yeast strain TMB3400 for furfural tolerance in glucose minimal media [56]. The evolved strain was then subjected to characterization by transcriptome and metabolic flux analysis. The studies showed that conversion of furfural to the less-toxic furfuryl alcohol was critical for furfural tolerance. At relatively low furfural concentrations (1.4 g/L), yeast primarily rely on an NADH-dependent furfural reductase. However, at concentrations of 2.4 g/L and greater, the role of NADPH-dependent enzymes becomes apparent due to increased metabolic flux through the NADPH-generating branch of the pentose phosphate pathway [56]. As opposed to the silencing of NADPH-dependent furfural reductase enzymes in the evolved *E. coli* strain, the evolved yeast strains actually had increased expression of NADPH-dependent furfural reductase enzymes, including ADH7 [56]. Thus, in yeast, removal of furfural is worth the cost of NADPH depletion. The reason for this difference from *E. coli* remains unclear, though it could possibly be attributed to the use of glucose as carbon source in the yeast studies and xylose in the *E. coli* studies, where glucose can be used to generate NADPH in the pentose phosphate pathway.

In an interesting contrast to the evolution for furfural tolerance, a study by Gorsich *et al.* [57] studied a gene disruption library of yeast strain BY4741 for mutations that increased furfural and 5-HMF sensitivity. This study revealed that the pentose phosphate pathway genes *zwf1*, *gnd1*, *rpe1* and *tkl1* were all associated with increased sensitivity. Subsequent overexpression of *zwf1* enabled tolerance of up to 4.8 g/L furfural, a concentration that is completely inhibitory to the parental strain [57]. These results yet again underscore the importance of NADPH, and the pathways that generate it, to furfural tolerance.

Clostridium

While the impact of inhibitory furans on butanologenic organisms is not as well characterized as with ethanologenic organisms, there are reports that butanol-producing *Clostridium acetobutylicum* is sensitive to fufural and 5-HMF and reduces them to furfuryl alcohol and 2,5-bis-hydroxymethfuran, respectively [58].

Conclusion

Here we have discussed the release and formation of aldehydes during the processing of biomass for the production of biorenewable fuels and chemicals. These aldehydes are concerning because they can limit growth, metabolism and product formation by the microbial biocatalysts. This problem can be addressed by chemical, biological or physical pretreatment of the biomass hydrolysate or pyrolysate to remove the aldehydes or by increasing the robustness of the biocatalyst. Biocatalyst robustness can be increased either through evolutionary enrichment processes or by rational engineering that specifically addresses the mechanism of inhibition.

In order to increase the economic viability of the production of biorenewable fuels and chemicals, this biocatalyst inhibition is a problem that needs to be addressed. Advancement is needed in both pretreatment methods to remove aldehydes and in increasing biocatalyst robustness. Though it should be noted that aldehydes are not the only problematic compounds present in the hydrolysate and pyrolysate – organic acids and methoxyphenols are concerning as well. In this manner, model aldehydes, such as furfural, provide an excellent example for how to address these inhibitory compounds.

Acknowledgments

This work was supported in part by the National Science Foundation Energy for Sustainability program (CBET-1133319), Center for Biorenewable Chemicals (EEC-0813570) and EPSCoR (EPS-1101284), the Iowa Energy Center (12-06) and the Iowa State University Bioeconomy Institute and Center for Sustainable Environmental Technologies. We thank undergraduate researcher Winston "Alex" Rosinger, part of the NSF BioMAP REU program, for his contribution in obtaining data for Table 3.

References

[1] Granda, C. B., L. Zhu and M. T. Holtzapple, Sustainable liquid biofuels and their environmental impact. *Environmental Progress*, 2007. 26(3): p. 233-250.

[2] Hoekman, S. K., Biofuels in the US - Challenges and Opportunities. *Renewable Energy*, 2009. 34(1): p. 14-22.

[3] Spiertz, J. H. J. and F. Ewert, Crop production and resource use to meet the growing demand for food, feed and fuel: opportunities and constraints. *Njas-Wageningen Journal of Life Sciences*, 2009. 56(4): p. 281-300.

[4] Hill, J., et al.., *Climate change and health costs of air emissions from biofuels and gasoline. Proceedings of the National Academy of Sciences of the United States of America*, 2009. 106(6): p. 2077-2082.

[5] Valentine, J., et al.., *Food vs. fuel: the use of land for lignocellulosic next generation' energy crops that minimize competition with primary food production. Global Change Biology Bioenergy*, 2012. 4(1): p. 1-19.

[6] Clomburg, J. M. and R. Gonzalez, Biofuel production in *Escherichia coli*: the role of metabolic engineering and synthetic biology. *Applied Microbiology and Biotechnology*, 2010. 86(2): p. 419-434.

[7] Du, J., Z. Shao and H. Zhao, *Engineering microbial factories for synthesis of value-added products. Journal of Industrial Microbiology and Biotechnology,* 2011. 38(8): p. 873-890.

[8] Halford, N. G., et al.., Sugars in crop plants. *Annals of Applied Biology*, 2011. 158(1): p. 1-25.

[9] Alvira, P., et al.., Pretreatment technologies for an efficient bioethanol production process based on enzymatic hydrolysis: A review. *Bioresource Technology*, 2010. 101(13): p. 4851-4861.

[10] Blanch, H. W., Bioprocessing for biofuels. *Current Opinion in Biotechnology*, 2012. 23(3): p. 390-395.

[11] Chundawat, S. P. S., et al.., Deconstruction of Lignocellulosic Biomass to Fuels and Chemicals, In: *Annual Review of Chemical and Biomolecular Engineering,* Vol. 2, J. M. Prausnitz, Editor. 2011. p. 121-145.

[12] Jarboe, L. R., et al.., Hybrid thermochemical processing: fermentation of pyrolysis-derived bio-oil. *Applied Microbiology and Biotechnology*, 2011. 91(6): p. 1519-1523.

[13] Vancov, T., et al.., Use of ionic liquids in converting lignocellulosic material to biofuels. *Renewable Energy*, 2012. 45: p. 1-6.

[14] Mills, T. Y., N. R. Sandoval and R. T. Gill, Cellulosic hydrolysate toxicity and tolerance mechanisms in *Escherichia coli. Biotechnology for Biofuels,* 2009. 2.

[15] Lee, S., A. K. Monnappa and R. J. Mitchell, *Biological activities of lignin hydrolysate-related compounds. Bmb Reports*, 2012. 45(5): p. 265-274.

[16] Zaldivar, J., A. Martinez and L. O. Ingram, Effect of selected aldehydes on the growth and fermentation of ethanologenic *Escherichia coli. Biotechnology and Bioengineering*, 1999. 65(1): p. 24-33.

[17] Zaldivar, J. and L. O. Ingram, Effect of organic acids on the growth and fermentation of ethanologenic *Escherichia coli* LY01. *Biotechnology and Bioengineering*, 1999. 66(4): p. 203-210.

[18] Dias, A. S., M. Pillinger and A. A. Valente, Dehydration of xylose into furfural over micro-mesoporous sulfonic acid catalysts. *Journal of Catalysis,* 2005. 229(2): p. 414-423.

[19] Eckert, C., et al.., Tunable solvents for fine chemicals from the biorefinery. *Green Chemistry*, 2007. 9(6): p. 545-548.

[20] Wu, G. X., M. Heitz and E. Chornet, Improved alkaline oxidation process for the production of aldehydes (vanillin and syringaldehyde) from steam-explosion of

hardwood lignin. *Industrial and Engineering Chemistry Research,* 1994. 33(3): p. 718-723.

[21] Xiang, Q. and Y. Y. Lee, Production of oxychemicals from precipitated hardwood lignin. *Applied Biochemistry and Biotechnology*, 2001. 91-3: p. 71-80.

[22] Taherzadeh, M. J., et al.., Characterization and fermentation of dilute-acid hydrolyzates from wood. *Industrial and Engineering Chemistry Research,* 1997. 36(11): p. 4659-4665.

[23] Patwardhan, P., Satrio, J., Brown, R., and Shanks, B., Influence of inorganic salts on the primary pyrolysis products of cellulose. *Bioresource Technology*, 2010. 101: p. 4646-4655.

[24] Guo, W., et al.., Performances of *Lactobacillus brevis* for Producing Lactic Acid from Hydrolysate of Lignocellulosics. *Applied Biochemistry and Biotechnology*, 2010. 161(1-8): p. 124-136.

[25] Qureshi, N., et al.., Effect of cellulosic sugar degradation products (furfural and hydroxymethyl furfural) on acetone-butanol-ethanol (ABE) fermentation using *Clostridium beijerinckii* P260. *Food and Bioproducts Processing,* 2012. 90(3): p. 533-540.

[26] Zhao, X., et al.., Effects of some inhibitors on the growth and lipid accumulation of oleaginous yeast *Rhodosporidium toruloides* and preparation of biodiesel by enzymatic transesterification of the lipid. *Bioprocess and Biosystems Engineering*, 2012. 35(6): p. 993-1004.

[27] Miller, E. N., et al.., Furfural Inhibits Growth by Limiting Sulfur Assimilation in Ethanologenic *Escherichia coli* Strain LY180. *Applied and Environmental Microbiology*, 2009. 75(19): p. 6132-6141.

[28] Palmqvist, E. and B. Hahn-Hagerdal, Fermentation of lignocellulosic hydrolysates. I: inhibition and detoxification. *Bioresource Technology*, 2000. 74(1): p. 17-24.

[29] Klinke, H. B., A. B. Thomsen and B. K. Ahring, Inhibition of ethanol-producing yeast and bacteria by degradation products produced during pre-treatment of biomass. *Applied Microbiology and Biotechnology*, 2004. 66(1): p. 10-26.

[30] Mussatto, S. I. and I. C. Roberto, Alternatives for detoxification of diluted-acid lignocellulosic hydrolyzates for use in fermentative processes: a review. *Bioresource Technology*, 2004. 93(1): p. 1-10.

[31] Purwadi, R., C. Niklasson and M. J. Taherzadeh, Kinetic study of detoxification of dilute-acid hydrolyzates by Ca(OH)(2). *Journal of Biotechnology,* 2004. 114(1-2): p. 187-198.

[32] Persson, P., et al.., Effect of different forms of alkali treatment on specific fermentation inhibitors and on the fermentability of lignocellulose hydrolysates for production of fuel ethanol. *Journal of Agricultural and Food Chemistry*, 2002. 50(19): p. 5318-5325.

[33] Martinez, A., et al.., Effects of Ca(OH)(2) treatments ("overliming") on the composition and toxicity of bagasse hemicellulose hydrolysates. *Biotechnology and Bioengineering*, 2000. 69(5): p. 526-536.

[34] Sirim, D., et al.., The Laccase Engineering Database: a classification and analysis system for laccases and related multicopper oxidases. Database: *the journal of biological databases and curation*, 2011. 2011: p. bar006.

[35] Zhang, T., N. Du and T. Tan, Biobutanol Production from Sweet Sorghum Bagasse. *Journal of Biobased Materials and Bioenergy*, 2011. 5(3): p. 331-336.

[36] Jonsson, L. J., et al.., Detoxification of wood hydrolysates with laccase and peroxidase from the white-rot fungus *Trametes versicolor. Applied Microbiology and Biotechnology,* 1998. 49(6): p. 691-697.

[37] Larsson, S., et al.., Comparison of different methods for the detoxification of lignocellulose hydrolyzates of spruce. *Applied Biochemistry and Biotechnology*, 1999. 77-9: p. 91-103.

[38] Wilson, J. J., L. Deschatelets and N. K. Nishikawa, Comparative fermentability of enzymatic and acid hydrolysates of steam-pretreated aspenwood hemicellulose by *Pichia stipitis* CBS-5776. *Applied Microbiology and Biotechnology*, 1989. 31(5-6): p. 592-596.

[39] Carter, B., P. C. Gilcrease and T. J. Menkhaus, Removal and Recovery of Furfural, 5-Hydroxymethylfurfural, and Acetic Acid From Aqueous Solutions Using a Soluble Polyelectrolyte. *Biotechnology and Bioengineering*, 2011. 108(9): p. 2046-2052.

[40] Chan, J. K. S. and S. J. B. Duff, Methods for mitigation of bio-oil extract toxicity. *Bioresource Technology*, 2010. 101(10): p. 3755-3759.

[41] Jarboe, L. R., P. Liu and L. A. Royce, Engineering inhibitor tolerance for the production of biorenewable fuels and chemicals. *Current Opinion in Chemical Engineering*, 2011. 1(1): p. 38-42.

[42] Jarboe, L. R., et al.., Development of ethanologenic bacteria, In: *Biofuels*, L. Olsson, Editor. 2007. p. 237-261.

[43] Yomano, L. P., et al.., Re-engineering *Escherichia coli* for ethanol production. *Biotechnology Letters*, 2008. 30(12): p. 2097-2103.

[44] Miller, E. N., et al.., Silencing of NADPH-Dependent Oxidoreductase Genes (*yqhD* and *dkgA*) in Furfural-Resistant Ethanologenic *Escherichia coli. Applied and Environmental Microbiology*, 2009. 75(13): p. 4315-4323.

[45] Gutierrez, T., L. O. Ingram and J. F. Preston, Purification and characterization of a furfural reductase (FFR) from *Escherichia coli* strain LYO1 - An enzyme important in the detoxification of furfural during ethanol production. *Journal of Biotechnology*, 2006. 121(2): p. 154-164.

[46] Nieves, I. U., et al.., Effect of reduced sulfur compounds on the fermentation of phosphoric acid pretreated sugarcane bagasse by ethanologenic *Escherichia coli. Bioresource Technology*, 2011. 102(8): p. 5145-5152.

[47] Bizouarn, T., et al.., Proton translocating nicotinamide nucleotide transhydrogenase from *E. coli*. Mechanism of action deduced from its structural and catalytic properties. *Biochimica Et Biophysica Acta-Bioenergetics,* 2000. 1457(3): p. 211-228.

[48] Jarboe, L. R., YqhD: a broad-substrate range aldehyde reductase with various applications in production of biorenewable fuels and chemicals. *Applied Microbiology and Biotechnology*, 2011. 89(2): p. 249-257.

[49] Turner, P. C., et al.., YqhC regulates transcription of the adjacent Escherichia coli genes *yqhD* and *dkgA* that are involved in furfural tolerance. *Journal of Industrial Microbiology and Biotechnology*, 2011. 38(3): p. 431-439.

[50] Wang, X., et al.., Increased Furan Tolerance in *Escherichia coli* Due to a Cryptic *ucpA* Gene. *Applied and Environmental Microbiology*, 2012. 78(7): p. 2452-2455.

[51] Khan, Q. A., F. A. Shamsi and S. M. Hadi, Mutagenicity of furfural in plasmid DNA. *Cancer Letters*, 1995. 89(1): p. 95-99.

[52] Uddin, S. and S. M. Hadi, Reactions of furfural and methylfurfural with DNA. *Biochemistry and Molecular Biology International*, 1995. 35(1): p. 185-195.

[53] Zheng, H., et al.., Increase in Furfural Tolerance in Ethanologenic *Escherichia coli* LY180 by Plasmid-Based Expression of *thyA*. *Applied and Environmental Microbiology,* 2012. 78(12): p. 4346-4352.

[54] Belfort, M., et al.., Primary structure of the *Escherichia coli thyA* gene ands its thymidylate synthase product. *Proceedings of the National Academy of Sciences of the United States of America-Biological Sciences,* 1983. 80(16): p. 4914-4918.

[55] Miller, E. N., et al.., Genetic changes that increase 5-hydroxymethyl furfural resistance in ethanol-producing *Escherichia coli* LY180. *Biotechnology Letters*, 2010. 32(5): p. 661-667.

[56] Heer, D., D. Heine and U. Sauer, Resistance of *Saccharomyces cerevisiae* to High Concentrations of Furfural Is Based on NADPH-Dependent Reduction by at Least Two Oxireductases. *Applied and Environmental Microbiology,* 2009. 75(24): p. 7631-7638.

[57] Gorsich, S. W., et al.., Tolerance to furfural-induced stress is associated with pentose phosphate pathway genes ZWF1, GND1, RPE1, and TKL1 in *Saccharomyces cerevisiae*. *Applied Microbiology and Biotechnology*, 2006. 71(3): p. 339-349.

[58] Zhang, Y., B. Han and T.C. Ezeji, Biotransformation of furfural and 5-hydroxymethyl furfural (HMF) by *Clostridium acetobutylicum* ATCC 824 during butanol fermentation. *New Biotechnology*, 2012. 29(3): p. 345-351.

[59] Chen, S. F., et al.., High-performance liquid chromatography method for simultaneous determination of aliphatic acid, aromatic acid and neutral degradation products in biomass pretreatment hydrolysates. *Journal of Chromatography A*, 2006. 1104(1-2): p. 54-61.

[60] Hodge, D. B., et al.., Soluble and insoluble solids contributions to high-solids enzymatic hydrolysis of lignocellulose. *Bioresource Technology*, 2008. 99(18): p. 8940-8948.

[61] Klinke, H. B., et al.., Characterization of degradation products from alkaline wet oxidation of wheat straw. *Bioresource Technology*, 2002. 82(1): p. 15-26.

[62] Alriksson, B., A. Cavka and L. J. Jonsson, Improving the fermentability of enzymatic hydrolysates of lignocellulose through chemical in-situ detoxification with reducing agents. *Bioresource Technology*, 2011. 102(2): p. 1254-1263.

[63] Martin, C., et al.., Ethanol production from enzymatic hydrolysates of sugarcane bagasse using recombinant xylose-utilising *Saccharomyces cerevisiae*. *Enzyme and Microbial Technology*, 2002. 31(3): p. 274-282.

[64] Carter, B., et al.., Detoxification of a Lignocellulosic Biomass Slurry by Soluble Polyelectrolyte Adsorption for Improved Fermentation Efficiency. *Biotechnology and Bioengineering*, 2011. 108(9): p. 2053-2060.

[65] Patwardhan, P. R., et al.., Distinguishing primary and secondary reactions of cellulose pyrolysis. *Bioresource Technology*, 2011. 102(8): p. 5265-5269.

[66] Demirbas, A., The influence of temperature on the yields of compounds existing in bio-oils obtained from biomass samples via pyrolysis. *Fuel Processing Technology*, 2007. 88(6): p. 591-597.

[67] Jung, S.-H., B.-S. Kang and J.-S. Kim, Production of bio-oil from rice straw and bamboo sawdust under various reaction conditions in a fast pyrolysis plant equipped

with a fluidized bed and a char separation system. *Journal of Analytical and Applied Pyrolysis*, 2008. 82(2): p. 240-247.

[68] Lian, J., et al.., Separation, hydrolysis and fermentation of pyrolytic sugars to produce ethanol and lipids. *Bioresource Technology*, 2010. 101(24): p. 9688-9699.

[69] Luo, Z. Y., et al.., Research on biomass fast pyrolysis for liquid fuel. *Biomass and Bioenergy*, 2004. 26(5): p. 455-462.

[70] Vitasari, C. R., G. W. Meindersma and A. B. de Haan, Laboratory scale conceptual process development for the isolation of renewable glycolaldehyde from pyrolysis oil to produce fermentation feedstock. *Green Chemistry*, 2012. 14(2): p. 321-325.

[71] Delgenes, J. P., R. Moletta and J. M. Navarro, Effects of lignocellulose degradation products on ethanol fermentations of glucose and xylose by *Saccharomyces cerevisiae, Zymomonas mobilis, Pichia stipitis*, and *Candida shehatae*. *Enzyme and Microbial Technology*, 1996. 19(3): p. 220-225.

[72] Oliva, J. M., et al.., Effect of lignocellulosic degradation compounds from steam explosion pretreatment on ethanol fermentation by thermotolerant yeast *Kluyveromyces marxianus*. *Applied Biochemistry and Biotechnology,* 2003. 105: p. 141-153.

[73] Sakai, S., et al.., Effect of lignocellulose-derived inhibitors on growth of and ethanol production by growth-arrested *Corynebacterium glutamicum* R. *Applied and Environmental Microbiology*, 2007. 73(7): p. 2349-2353.

[74] Jayakody, L. N., N. Hayashi and H. Kitagaki, Identification of glycolaldehyde as the key inhibitor of bioethanol fermentation by yeast and genome-wide analysis of its toxicity. *Biotechnology Letters*, 2011. 33(2): p. 285-292.

[75] Ranatunga, T. D., et al.., The effect of overliming on the toxicity of dilute acid pretreated lignocellulosics: the role of inorganics, uronic acids and ether-soluble organics. *Enzyme and Microbial Technology*, 2000. 27(3-5): p. 240-247.

[76] Converti, A., et al.., Wood hydrolysis and hydrolysate detoxification for subsequent xylitol production. *Chemical Engineering and Technology*, 2000. 23(11): p. 1013-1020.

[77] Weil, J. R., et al.., Removal of fermentation inhibitors formed during pretreatment of biomass by polymeric adsorbents. *Industrial and Engineering Chemistry Research*, 2002. 41(24): p. 6132-6138.

In: New Developments in Aldehydes Research
Editors: L. Torrioni and E. Pescasseroli

ISBN: 978-1-62417-090-4

Chapter 5

CO-OXIDATION PROCESSES PROMOTED BY *N*-HYDROXYPHTHALIMIDE/ ALDEHYDE SYSTEM

Lucio Melone and Carlo Punta*

Department of Chemistry, Materials, and Chemical Engineering
"G. Natta" Politecnico di Milano, Milano, Italy

ABSTRACT

Aldehyde autoxidation rates are very high even at room temperature and atmospheric pressure and this property has been often used to promote the so-called "co-oxidation" processes, in which a mixture of an aldehyde and another less reactive organic substrate is submitted to molecular oxygen. The present work is addressed to overcome the most recent developments in this field, covering the last decade.

After a brief overview on significative co-oxidative processes mediated by aldehydes, we focus on the recent results reported by the authors for the selective epoxidation of olefins and the selective oxidation of alkylaromatic to the corresponding hydroperoxides promoted by the new *N*-hydroxyphthalimide (NHPI)/aldehyde catalytic system under mild aerobic oxidations. Enthalpic, polar, solvent and temperature effects are widely disclosed. Moreover, the free-radical reaction mechanism based on the in situ generation of phthalimido-*N*-oxyl (PINO) radical by means of the "molecule-induced homolysis" of NHPI is deeply investigated and supported by experimental evidences.

1. INTRODUCTION

The selective aerobic oxidation of hydrocarbons is generally recognized by the scientific and industrial community as one of the most convenient procedures for the introduction of functional groups onto organic feedstock, leading to the synthesis of more complex molecules

* Department of Chemistry, Materials, and Chemical Engineering "G. Natta" Politecnico di Milano, Via Leonardo da Vinci 32, I-20131 Milano, Italy.

with high added value [1]. Moreover, the eco-friendly standards require oxidants to be able to combine a low environmental impact with an economical convenience, operating under mild conditions, with inexpensive reagents and high levels of atom economy. At least in principle, molecular oxygen seems to be the ideal oxidant from this point of view.

However, it is well known how the use of oxygen strictly depends upon the employment of activating systems, especially when high temperatures are detrimental for the selectivity of the process. The careful choice of the activation protocol is then crucial not only for the selective achievement of the desired products, but also to remain in accordance with the commonly recognized rules of green-chemistry.

Sacrificial reductants have been widely reported in the literature as reactive agents able to promote the autoxidation reaction of less reactive hydrocarbons.

In this contest, oxygen activation using an aldehyde as a final sacrificial reagent has attracted increasing attention since the very first protocol reported in 1926 [2]. Aldehydes are exceptions among the majority of organic compounds as their autoxidation rates are very high even at room temperature and atmospheric pressure.

This property has been used in several co-oxidation processes, according to which, in a mixture of an aldehyde and another organic substrate, the autoxidation of the aldehyde is able to promote in turn the oxidation of the less reactive partner.

This approach has found fundamental applications, especially in the field of selective epoxidation of olefins, thanks to the pioneering works of Ikawa [3], who contributed to a better comprehension of the radical reaction mechanism, by determining the kinetics at room temperature of the oxidation of cyclohexene and 2-butene (Scheme 1).

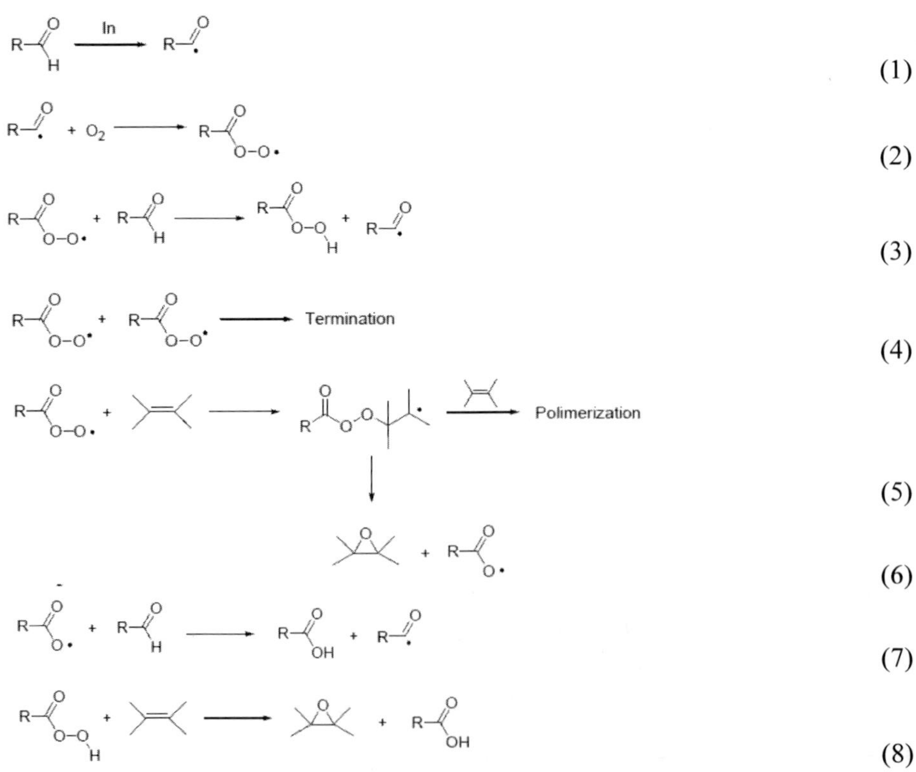

Scheme 1. Free-radical co-oxidation mechanism in the presence of O_2, aldehydes and olefins.

The activating effect of aldehydes results from the low R (O=C)-H bond dissociation energy of the acyl group, especially in the liquid phase (Equation 1), and the concomitant relatively high stability of the peroxo radicals which are immediately formed in the presence of O_2 (Equation 2). These peroxo radicals can act as initiators and terminators of radical chains, influencing the selectivity of the overall process [4].

Almost thirty years later, Mukaiyama [5] proposed a series of oxidation systems in combination with metallo-complexes of Ni^{2+}, Co^{2+}, and Fe^{2+}, including chiral catalysts [6].

More recently, the practical interest for this kind of reaction has led to the development of heterogeneous catalysts [7], supporting the complexes also on inorganic matrices [8], while branched aliphatic aldehydes such as isobutyraldehyde were shown to be the most suitable for these processes.

The aerobic co-oxidation promoted by aldehydes could be considered a non-green and expensive process, according to what claimed before, as it requires the use of a sacrificial reagent.

Nevertheless, there are three different circumstances that would make the oxygen activation using an aldehyde competitive for practical application:

I. The by-product of aldehyde oxidation, that is the carboxylic acid, is an interesting compound of commercial relevance;
II. The high efficiency in substrate conversion, the high selectivity in the final product, and/or the field of application, together with the low cost and environmental impact of the selected aldehyde, are so significative to justify the reagent sacrifice;
III. The aldehyde is not used as a stoichiometric reagent, but in tiny amounts as initiator of free-radical chains.

In Section 2 we report a selection of the most significative protocols reported in the literature along the last decade, showing the ongoing interest on the aerobic oxidation activated by aldehydes, and the implementation of some of these processes for larger productions.

In Section 3 we describe in detail our recent contributions in the field, discussing the results achieved by combining the co-oxidation of aldehydes in the presence of molecular oxygen with the catalytic activity of *N*-hydroxyphthalimide (NHPI).

2. Oxygen Activation Using Sacrificial Aldehydes

The use of aldehydes as sacrificial reductive activators has found ample use in the field of selective epoxidation of olefins, due both to the importance of these synthetic intermediates and to the economical and environmentally friendly nature of the approach, which allows to operate with molecular oxygen as principal oxidant.

Besides this main application, sacrificial aldehydes have been also employed in the aerobic oxidation of sulphur containing compounds, both in organic synthesis and as a route for the oxidative desulfurization of transportation fuels, as well as in the selective conversion of alcohols and hydrocarbons.

2.1. Epoxidation of Olefins

Following the early studies by Mukaiyama [5], many metal catalysts have been reported showing highly catalytic performance for the aerobic epoxidation in the presence of aldehydes.

In this contest, the mediation of metalloporphyrins in promoting free radical processes has been widely investigated. Indeed, these derivatives (as well as other metallo-complexes) are improperly defined catalysts, as they essentially behave as initiators of free radical chains, being involved in the fast acyl radical generation (Scheme 1, Equation 1) according to the generic Equation 9, while they do not intervene in the propagation phase [9].

It was demonstrated that the intervention of free-radical species in these co-oxidative processes is responsible for the macrocyclic destruction [10]. The degradation is minimized by addition of the olefin and oxygen, while it is really faster when less robust metalloporphyrins are used [9].

For these reasons, many efforts have been recently devoted for the development of metalloporphyrins to be able to combine their low cost with a high turnover number of the catalyst and high resistance to degradation. The incorporation of electron-withdrawing substituents at the β- and/or meso-position has been found to be a winning strategy to increase stability and activity of metalloporphyrins to be employed in oxygenation reactions [11].

In 2007 Ji and co-workers reported how ppm amounts of manganese *meso*-tetraphenylporphyrin 1 (Figure 1) were able to promote a highly efficient epoxidation of olefins in the presence of oxygen by using a substrate-to-isobutyraldehyde molar ratio of 1:5 [12]. This catalyst proved to have an extremely high turnover number, with values up to 700 millions, that is comparable with enzymatic-like catalysis, and allowed to perform larger scale epoxidation processes of cyclohexene (20 mmol) with isolated yields higher then 90 %. A couple of years later [13], a remarkable enhancement of reactivity was achieved by the same research group by employing for the first time a μ-oxo-bisiron(III) tetraphenylporphyrin 2 (Figure 1) for the aerobic oxidation of cyclohexene. In this case the turnover number was even double (1400 millions) compared with the previously reported protocol.

$$R{-}C(=O)H + M^{n+} \longrightarrow R{-}\dot{C}{=}O + M^{(n-1)+} + H^{+} \qquad (9)$$

Among the several electron-withdrawing groups introduced to enhance porphyrins' stability, perfluoroalkyl chains were reported to be particularly advantageous. They are inert under aerobic conditions and they contribute to prevent the oxidation of the macrocycle by stabilizing the HOMO of the complex. Moreover, these derivatives show higher solubility also in fluorous media, favouring their employment in fluorous biphasic catalysis. In 2006 Chen and co-workers reported a fluorous biphasic catalytic oxidation of alkenes in good to excellent yields, promoted by (β-perfluoroalkylated tetraphenylporphyrin) cobalt complexes 3 (Figure 1) with 1 atm of air and in the presence of 2-methylpropanal as sacrificial reductant [14].

With this approach it was possible to realize an ease recyclability of the catalyst, as a reuse of the fluorous phase up to five times did not show any apparent loss of catalytic activity.

Besides porphyrins, other metallo-complexes were presented as effective catalysts for selective aldehyde-mediated epoxidations with dioxygen. For example, the recently reported manganese complex $Mn(HL)_2Cl_2$ 4 (where HL is the N-4'-methylphenyl-2-pyridinecarbox-amide ligand, Figure 2) showed a better solubility in the classical reaction medium (CH_2Cl_2 or $ClCH_2CH_2Cl$), with consequent increased catalytic activity and final selectivity in the desired epoxide [15]. However, if the main role of the metal complex relies in generating the acyl radical in the initiation step, the use of simple systems not involving organic complex molecules should be taken into account, especially for practical purposes.

Indeed, in 2011 Gonsalves et al. [9] clearly demonstrated that, by operating with a 10% amount of the very simple heterogeneous catalyst complexes, like $Mn(OAc)_2{\cdot}H_2O$ or $Co(OAc)_2$, and with a substrate-to-isobutyraldehyde molar ratio of 1:5, it was possible to achieve the complete conversion of citronellol to the corresponding epoxide in 6 h without affecting the terminal hydroxyl group.

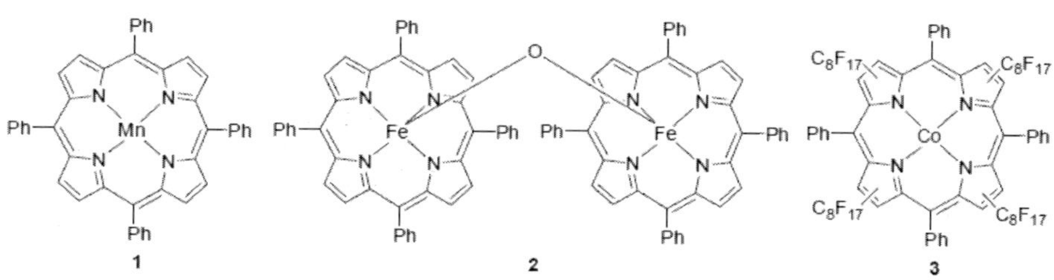

Figure 1. Significative examples of metalloporphyrins employed in the selective epoxidation of olefins with molecular oxygen in the presence of aldehydes.

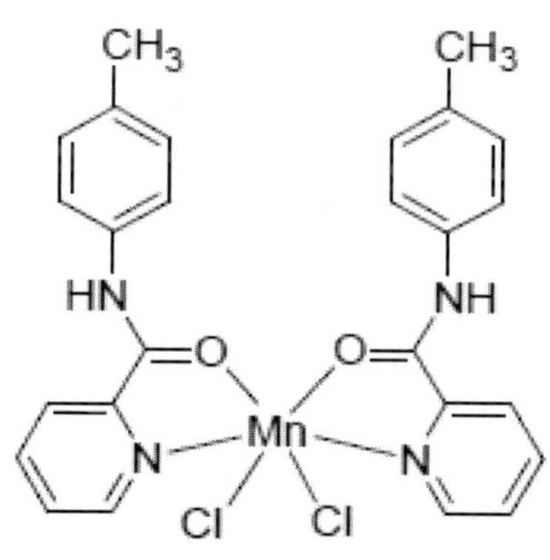

4

Figure 2. Manganese complex $Mn(HL)_2Cl_2$ 4.

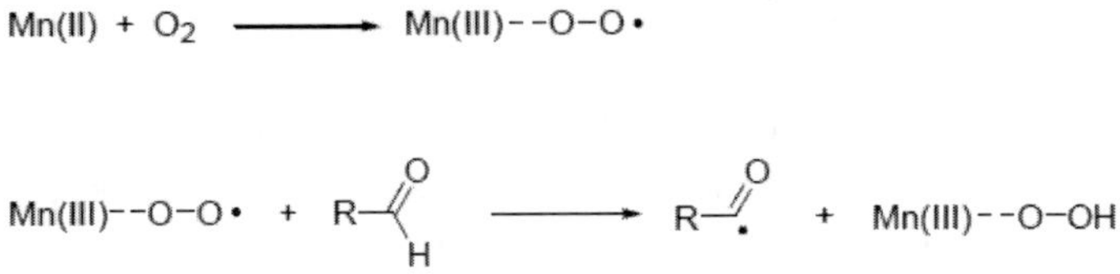

(10)

(11)

Scheme 2. Alternative mechanism for the formation of acyl radical in the from the corresponding aldehyde by means of Mn(II) in the presence of molecular oxygen.

The protocol, which was successfully extended to a wider range of olefins, could occur according to Equation 9 or by direct interaction of Mn(II) with O_2, forming a superoxide which in turn was responsible for acyl radical formation by hydrogen abstraction (Scheme 2).

In consideration of the fact that the predominant role of metal-complexes is expressed in the initiation phase, it should be pointed out that the use of metal catalysts was not mandatory to obtain high yields and selectivity in epoxides [16]. In fact, good results were obtained when a radical chain initiator was used [17].

This approach resulted of particular interest for the synthesis of epoxidized vegetable oils and fatty acid derivatives, which represent important building blocks for the production of epoxy resins, plasticizers and stabilizers.

It is well known that, in the absence of any additive or in the presence of metal-complexes, molecular oxygen attacks the allylic position of the unsaturated fatty acid, leading to degradation products such as aldehydes, ketones and acids via hydroperoxides formation [18].

In 2008 Köckritz et al. [19] suggested how to overcome this limitation, succeeding in the epoxidation of methyl oleate with molecular oxygen by operating in the presence of an excess isobutyraldehyde with azobisisobutyronitrile (AIBN) as radical chain initiator (Scheme 3).

2.2. Oxidation of Sulfides, Alcohols and Alkanes

Oxidation of sulfides finds important applications in organic synthesis for the production of sulfoxides intermediates. Among the several methods reported in the literature, the co-oxidation of aldehydes and organosulfur compounds with molecular oxygen, in the absence [20] or presence [21] of transition metal salts, was proposed as an alternative environmentally friendly route with respect to classical procedures, which often require expensive and non-green oxidants.

Once again, the proposed mechanism consists in the in situ generation of peroxy acid, which is in turn the real responsible of sulfide oxidation. This approach was successfully applied to bis-sulfides, performing their regioselective aerobic oxidation to monosulfoxides [22] with yields higher than 90 %.

In line with the increase of environmental concern, one of the main tasks in the oxidation of sulphur-containing compounds is represented by the development of advantageous oxidative protocols for the desulfurization of transportation fuels. The oxidative desulfurization consists of the selective oxidation of the oganosulfur compounds followed by the removal of the oxidized products from the treated fuel.

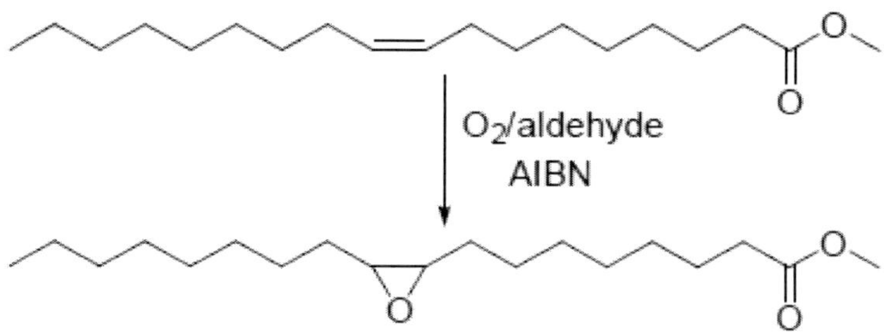

Scheme 3. Methyl oleate epoxidation in the presence of O_2/aldehyde system.

This approach, which represents an alternative to classical hydrodesulfurization, has several advantages such as the mild reaction conditions, the high selectivity, the lower economic impact and the potential application on a wider range of substrates, including sterically hindered sulfides.

Among the several procedures reported for this purpose, oxidation by means of peracids has attracted particular attention, due to the fact that these reactions rapidly occur at room temperature and atmospheric pressure [23]. Nevertheless, their use on large-scale and their storage is dangerous.

In 2004 Murata and co-workers [24] reported a new protocol to totally convert dibenzothiophene to the corresponding sulfone in just 15 minutes and under very mild conditions (40 °C and atmospheric pressure), based on the sacrificial oxidation of *n*-octanal with molecular oxygen promoted by $Co(OAc)_2$ or $CoCl_2$ salts (Scheme 4).

This method was successfully applied to ultra deep desulfuration of a commercial gas oil, consisting in the oxidation of the diesel fuel sample with O_2/octanal/Co(II) system and the consequent removal of sulfur-containing compounds via extraction with acetonitrile followed by adsorption with Al_2O_3. According to this approach, more than 97 % of sulfur could be removed from the oil. In 2001, Nobile and coworkers performed the selective oxidation of benzylic and secondary alcohols to the corresponding carbonylic compounds, by operating in the presence of a similar system consisting in O_2 and 2-methylpropanal, with catalytic amounts of $Co(acac)_2$ [25].

In this report, the authors demonstrated that the aerobic oxidation could be also promoted by a heterogeneous analogue of $Co(acac)_2$, consisting in a copolymer obtained by reaction of $Co(AAEMA)_2$ ($AAEMA^-$ = deprotonated form of 2-(acetoacetoxy)ethyl methacrylate) with *N,N*-dimethylformamide. This supported cobalt polymer behaved as an active catalyst, affording the desired products with yields > 95%, and ensuring recyclability with turnovers ranging between 56 and 59 h^{-1}.

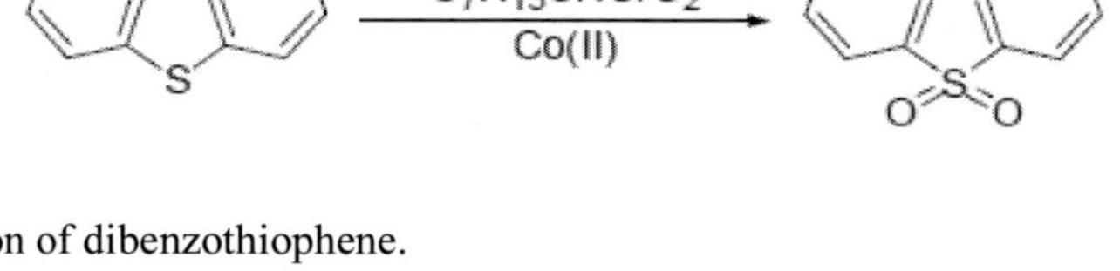

Scheme 4. Oxidation of dibenzothiophene.

Among the significant number of catalytic protocols for aerobic oxidation of alcohols in the presence of a sacrificial aldehyde and first row transition metal complexes with different ligand types, in 2001 Pedro and co-workers reported some interesting results by combining a dioxygen/pivalaldehyde system with square-planar anionic cobalt(III) complexes with the ligand o-phenylenebis(*N*'-methyloxamidate) ad its oxamate and bis(oxamate) derivatives (Figure 3) [26].

These catalysts resulted particularly effective in the oxidation of both aliphatic and benzylic alcohols. A deep investigation of the reaction mechanism suggested the intervention of high-valent oxo-cobalt (IV) species as key oxidising species. In consideration of the chemical composition of the worldwide oil, the selective oxidation of alkanes with O_2 has attracted a growing interest, due to the opportunity of producing important building blocks for further synthesis of organic derivatives with high added value. Thus, the sacrificial reduction of aldehyde in the presence of molecular oxygen has been widely used also to afford the selective oxidation of hydrocarbons.

Theyssen and Leitner [27, 28] recently investigated the applicability of this approach in compressed (supercritical) carbon dioxide medium, combining in a green process the use of molecular oxygen with the additional advantage of operating in a totally inert environment, which guarantees an easy recovery of the product and recyclability of the solvent. In these studies, acetaldehyde was chosen as cheap and CO_2-soluble oxygen activator, which was converted to acetic acid, while the initiation of the free-radical chain was ascribed to the stainless steel of the reactor walls (Scheme 5). For the oxidation of alkanes the choice of the type and amount of aldehyde can influence the selectivity of the overall process, as aldehydes can chemically interact with radical species. This aspect will be emphasized in Section 3.2.

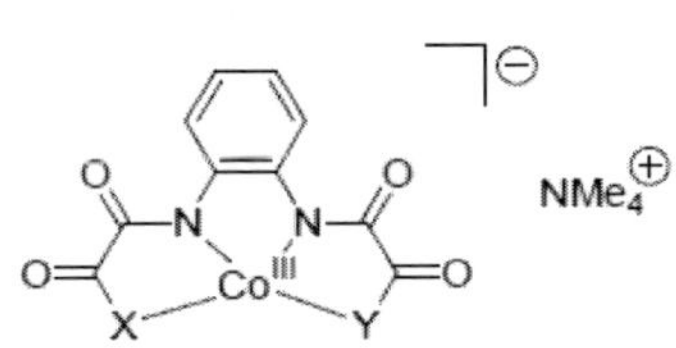

X = N(CH$_3$) Y = N(CH$_3$) **5**

X = N(CH$_3$) Y = O **6**

X = O Y = O **7**

Figure 3. Square-planar anionic cobalt(III) complexes.

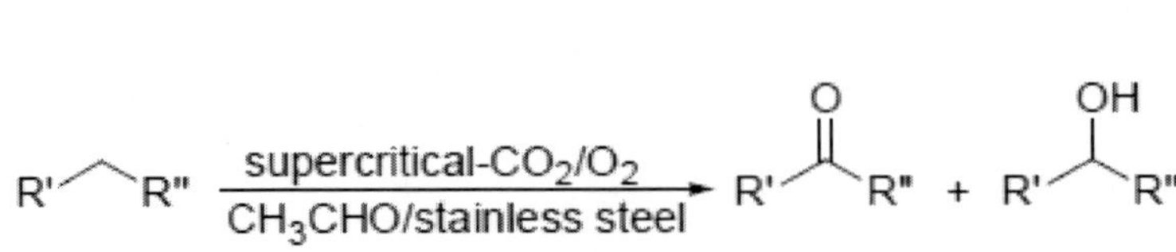

Scheme 5. Oxidation of hydrocarbons in supercritical CO_2.

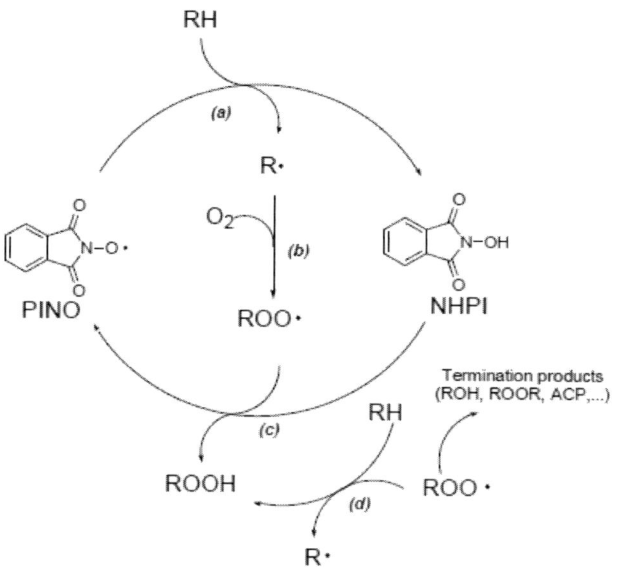

Scheme 6. Catalytic role of NHPI and PINO in autoxidative processes.

3. Selective Aerobic Oxidation Promoted by NHPI/Aldehyde Systems

In the last decade NHPI has gained a role of protagonist as an effective catalyst in the field of selective aerobic oxidations [29]. In these protocols, the phthalimido-*N*-oxyl (PINO) radical, generated in situ from NHPI by different initiators, promotes the C-H activation by hydrogen abstraction with higher efficiency with respect to the peroxyl radicals normally involved in autoxidation processes (Scheme 6).

Thanks to the strict collaboration with the research group of Professors Lucarini and Pedulli (University of Bologna), we have significantly contributed to the interpretation of the reaction mechanisms, also via thermodynamic and kinetic characterization of NHPI-systems [30]. In a classical autoxidation process, where the hydrogen abstraction reactions determine the efficiency of the propagation phase, the overall chain proceeds at a reasonable rate only if the bond dissociation enthalpy (BDE) of the bonds formed is higher or at least similar to the strength of the bonds cleaved. Thus, the catalytic role of NHPI had to be associated to the BDE of the O-H group. Lucarini et al. [30b] measured this value, which resulted to be 88.1 kcal/mol.

From one point of view, this result justified the activity of PINO radical, which could undergo thermoneutral or slightly endothermic hydrogen abstraction reactions from a wide range of organic compounds (path (a), Scheme 6), in contrast with other nitroxyl radicals like (2,2,6,6-Tetramethyl-piperidin-1-yl)oxyl (TEMPO) (the O-H BDE of the corresponding *N*-hydroxy derivative is 70 kcal/mol). On the other side, the BDE value was similar to the BDE of the O-H group in hydroperoxides, usually formed in autoxidative processes from the corresponding peroxyl radicals by means of hydrogen abstraction reactions (paths (b) and (d),

Scheme 6). Thus, this result suggested that the catalytic role of PINO could not be simply ascribed to enthalpic reasons. Indeed, an enhanced polar effect in oxidative processes catalyzed by NHPI was observed for the aerobic oxidation of substituted benzylic alcohols, and this behaviour was related to a more-pronounced electrophilic character of the PINO radical relative to the peroxyl radical, which induces a stabilization of the transition state (Scheme 7) [31]. As a consequence, PINO radical results to be always more reactive than peroxyl radicals in the hydrogen-abstraction step, as it is clearly evidenced by the ratio $k_{PINO}/k_{ROO\bullet}$ reported in Table 1.

Moreover, NHPI is also involved in a second step of the radical chain, acting as a good hydrogen atom donor in the fast trapping by of peroxyl radicals (path (c)), the latter formed according to path (b) (Scheme 6).

Lucarini et al. [29b] were able to determine the kinetic constant for Equation 12, which resulted to be relatively high ($k_{12} = 7.2 \bullet 10^3\ M^{-1}s^{-1}$) and even if the reaction is reversible, it is reasonable to assume it is shifted to the right, as estimated by a DFT calculation by Hermans et al. [35] ($k_{-12} = 32.4$).

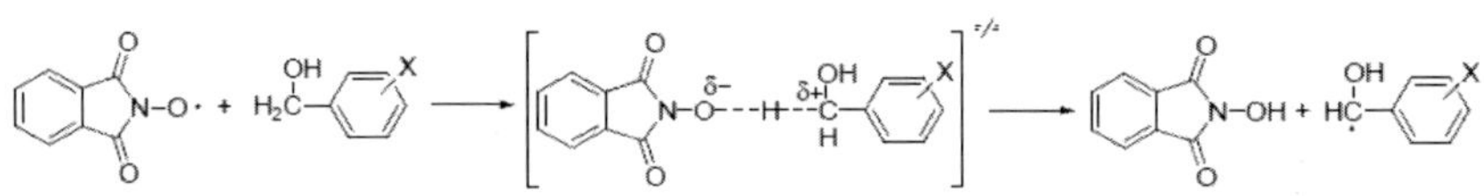

Scheme 7. Polar effect in hydrogen abstraction from benzyl alcohols by PINO.

Table 1. A comparison of kinetic constants for the hydrogen abstraction reaction from different organic substrates by peroxyl and PINO radicals, respectively

R-H	$k_{ROO\cdot}$ [a]	k_{PINO}	$k_{PINO}/k_{ROO\cdot}$
benzylalcohol	2.4	5.65[b] 14.1[c] 7.8[d]	2.4 5.9 10.2
Ph_2CH_2	1.05	6.63[b]	6.3
tetralin	1.6	10.8[d]	6.7
cumene	0.18	3.25[c]	18
ethylbenzene	0.65	1.12[c] 0.925[d]	1.72 1.42
1-methylnaphthalene		1.43[b]	
toluene	0.08	0.127[c]	1.6
tetrahydrofuran	1.1	0,72[d]	0.65

[a][32]; [b][33]; [c][29b]; [d][34].

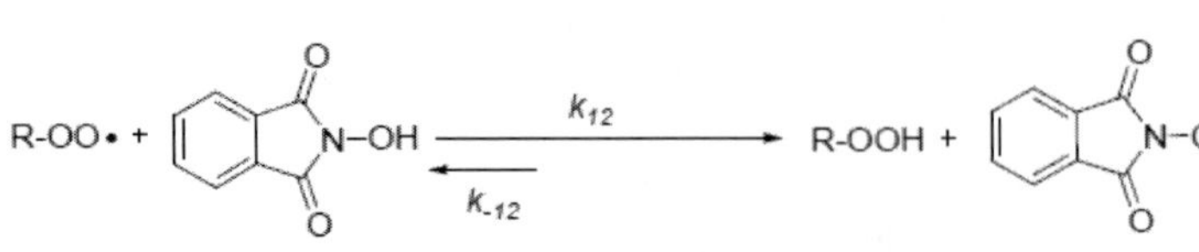

(12)

Thus, the role of NHPI in promoting autoxidative processes is dual: it catalyses the reaction intervening in path (a) and it limits termination reactions between peroxyl radicals, which lead to undesired products, by acting as good H-atom donor (path c). This deep mechanistic investigation allowed us the development of a large variety of new selective oxidation processes of potential industrial and synthetic interest, using molecular oxygen as final oxidant and a wide range of metal and non-metal initiators [29]. As evidenced before, the use of metal catalysts for the activation of aerobic co-oxidation reactions in the presence of sacrificial aldehydes is not mandatory, due to the unique role as radical chain initiators, and in few cases it is detrimental for the selectivity of the process. On the other side, the employment of initiators operating by thermal decomposition (such as AIBN and benzoyl peroxides) often require relatively high temperatures, which are not always compatible with the overall protocol. Co-oxidation processes carried out in our laboratories in the presence of NHPI/aldehydes systems allowed to generate PINO radical under mild, metal-free conditions, according to a molecule-induced homolysis mechanism. This allowed us to develop both the selective epoxidation of olefins and the hydroperoxidation of alkylaromatics.

3.1. Molecule-Induced Homolysis of NHPI

Almost 20 years ago, Minisci and co-workers investigated the oxidation of a wide range of organic compounds (including hydrocarbons, alcohols, ethers and aldehydes) by dioxiranes [36] and peracids [37], in order to disclose the reaction mechanism of this controversial reaction. In contrast with the widely accepted “concerted oxenoid oxygen insertion”, an ionic mechanism, they proposed a “molecule-induced homolysis” in the transition state, which leads to radical pairs by hydrogen abstraction (Scheme 8). The authors ascribed the driving force of this step to the high BDE of the O-H bonds formed in the products (i.e. water), so that the weaker is the R-H bond in the substrate, the more effective is the hydrogen abstraction. On the bases of these preliminary studies and in consideration of the relatively low value of the O-H BDE in NHPI (88.1 kcal/mol), in collaboration with the same group we suggested that an analogous homolysis induced by peracids could occur for this *N*-hydroxy derivative, leading to the formation of PINO radical under mild and metal-free conditions (Scheme 9) [38]. Experimental evidences supported this assumption. From a spectroscopic point of view, Lucarini et al. observed the characteristic EPR (Electron Paramagnetic Resonance) spectrum of PINO radical by simply adding NHPI to a solution of acetonitrile containing *m*-chloroberbenzoic acid (a(2H) = 0.46 G; a(N) = 4.77 G) [38].

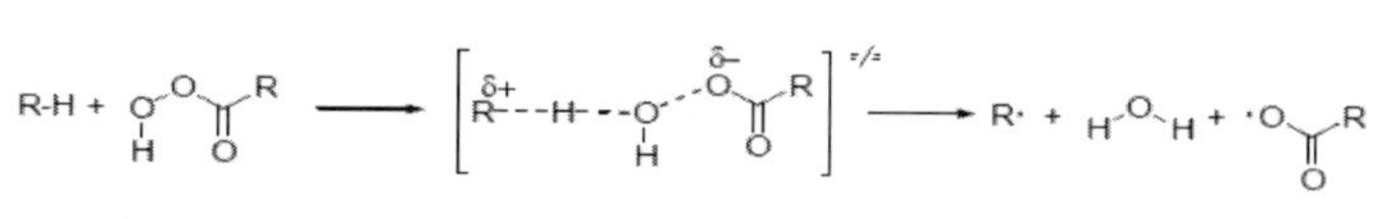

Scheme 8. Transition state for the oxidation of organic compounds by peracids.

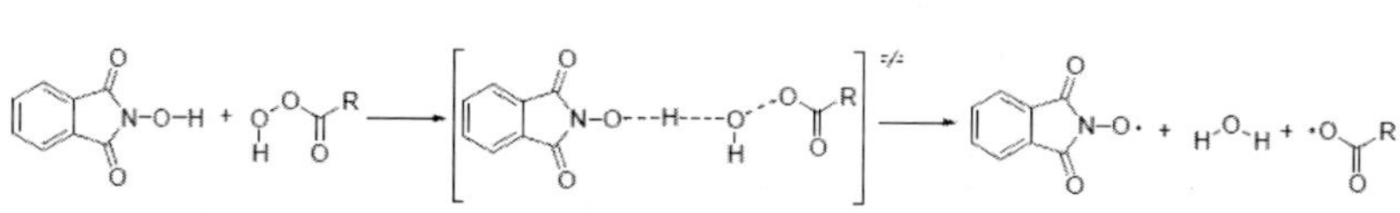

Scheme 9. Transition state for the “molecule-induced homolysis” of NHPI by peracids.

HPLC and GC analysis of the products of the same reaction revealed the formation of *m*-chlorobenzoic acid (90%) and Cl-benzene (10%) (Scheme 10) [38]. This distribution could be explained only owning the formation of the acyloxyl radical as described in Scheme 9.

Moreover, this hypothesis was also supported by the product distribution observed when the same reaction between *m*-chloroberbenzoic acid and NHPI was conducted in benzene as a solvent: in addition to *m*-chlorobenzoic acid, phenyl *m*-chlorobenzoate and *m*-chlorobiphenyl were also detected (Scheme 11)

On the bases of these experimental results, we suggested that the aerobic oxidation of aldehydes could be performed for the in situ generation of the corresponding peracids in the presence of NHPI, promoting co-oxidative processes catalyzed by PINO.

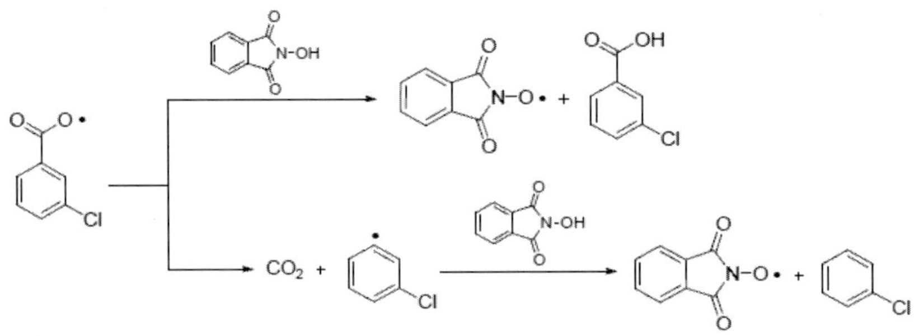

Scheme 10. Product distribution for the reaction between *m*-chloroberbenzoic acid and NHPI in acetonitrile.

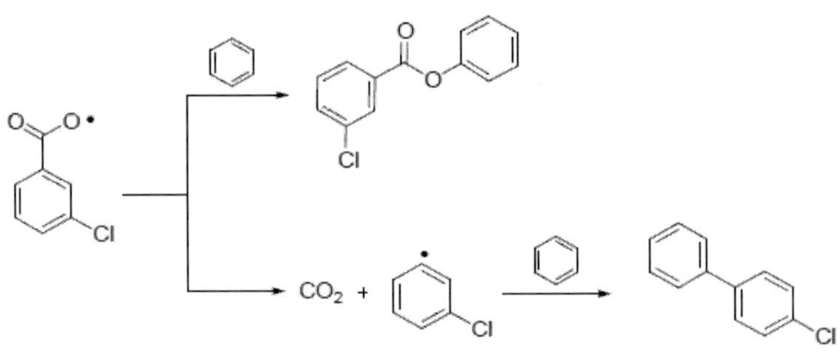

Scheme 11. By-product distribution for the reaction between *m*-chloroberbenzoic acid and NHPI in benzene.

3.2. Selective Epoxidation of Olefins with NHPI

In 2006 we reported the NHPI-catalyzed selective epoxidation of α-olefins and cyclic olefins with good yields, by means of aerobic co-oxidation of acetaldehyde (alkene-to-aldehyde molar ratio was 1:3) at room temperature and under atmospheric pressure of molecular oxygen (Scheme 2) [38].

This protocol was successfully applied on larger scale (1 liter Buchi glass vessel) for the synthesis of propylene oxide from the corresponding propene, with complete selectivity in the desired product [39].

Internal olefins resulted unreactive under these operative conditions. This selectivity, together with the stereochemistry observed for the epoxidation of limonene (67% of the *cis* isomer and 33% of the *trans* isomer) clearly indicated that peracetic acid would not intervene in the epoxidation step. In fact, with peracids, internal acyclic olefins are always more reactive than α-olefins (up to 300 times more reactive) and the stereochemistry of epoxidated limonene is completely different (41% of the *cis* isomer and 59% of the *trans* isomer).

Thus, we suggested an alternative free-radical mechanism according to which the acyl radical, generated by hydrogen abstraction from acetaldehyde by PINO (Equation 13), reacted with oxygen (Equation 14), affording the corresponding acyl peroxyl radical. The latter in turn adds to the olefin (Equation 15), generating a radical adduct which undergoes fast decomposition towards the desired product (Scheme 13). It is reasonable to assume that peracetic acid is actually formed in the reaction medium according to Equation 16, but its reaction with NHPI to promote the molecule induced homolysis is so fast that it does not compete with acyl peroxyl radical in the epoxidation of olefin.

The absence of reactivity for internal olefins can most probably be attributed to a simple steric effect, which inhibits the insertion of the acyl peroxyl radical on the double bond.

In spite of the significative environmental ad economical advantages introduced by Minisci epoxidation, the main limit associated to this process was due to the low relative efficiency owing to long batch reactor residence times (24–48 h).

In order to overcome this major drawback, we have recently reported the application of an analogous epoxidizing system under continuous flow conditions [40], by means of a new technology designed and developed by Bjørsvik and co-workers (Bergen University and Fluens Synthesis): the multijet oscillating disk (MJOD) reactor [41].

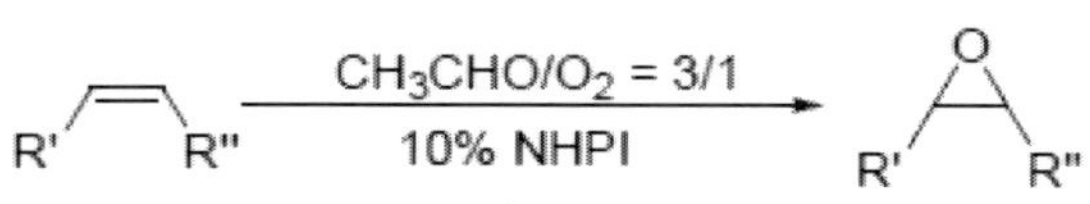

Scheme 12. Selective epoxidation of alkenes promoted by NHPI/acetaldehyde/O_2.

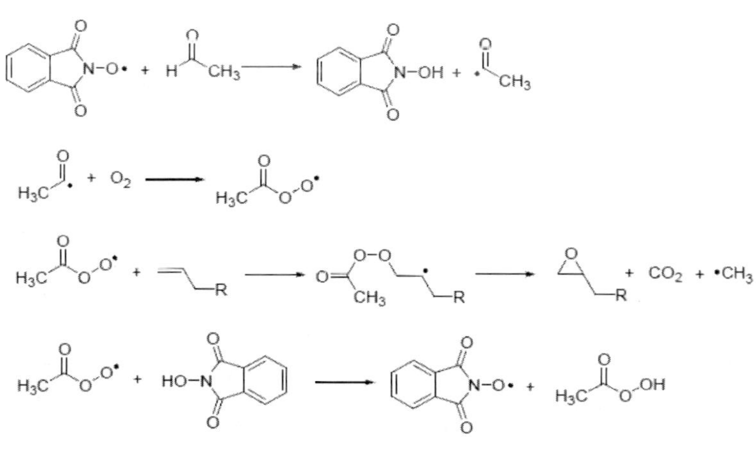

(13)

(14)

(15)

(16)

Scheme 13. Reaction mechanism.

Indeed, we succeeded in the epoxidation of a wide range of olefins, both by achieving high conversions and selectivity in the desired products and by accelerating the overall process, shortening the residence time with respect to the batch protocol up to only 1-4 h. The selectivity observed by operating under continuous flow conditions was somehow different. In fact, internal olefins were epoxidized as well. This result suggested that, in this case, peracetic acid was involved in the epoxidation process. The excellent mass transfer which can be achieved by using the MJOD reactor, probably led to an augmented production of peracid as well as an extended contact between the in situ produced peracid and the olefin.

3.3. Selective Hydroperoxidation of Secondary and Tertiary Alkylaromatics

The selective oxidation of hydrocarbons to the corresponding hydroperoxides represents one of the tasks in the filed of aerobic oxidations, due to the importance of these compounds as intermediates for the synthesis of a wide range of derivatives of industrial relevance.

Important examples are cumyl hydroperoxide (CH, an intermediate in the Hock process for the production of phenol) and phenyl ethyl hydroperoxide (PEHP, widely employed in the production of propylene oxide and styrene in the Shell SM/PO process), obtained from cumene (CU) and ethylbenzene (ETB), respectively.

In 1997, Einhorn and co-workers conducted a series of aerobic oxidations on a wide range of hydrocarbons [42], including CU and EB, combining stoichiometric amounts of acetaldehyde with catalytic quantities of NHPI. In spite of the high conversions reported in that protocol, no traces of hydroperoxides were reported (Table 2, runs 1 and 4).

On the bases of our previous studies on the reactivity of NHPI in the presence of acetaldehyde, we decided to investigate in detail these reactions involving both CU [43] and ETB [44], as we were expected to find a different selectivity for these substrates.

In fact, by operating with stoichiometric amounts of acetaldehyde, the oxidation of CU afforded CH as the main product, being cumyl alcohol (CA) the principal by-product, while only tiny amounts of acetophenone (AP) were observed (Table 2, run 2).

Similarly, PEHP resulted to be the predominant product deriving from the oxidation of ETB, whereas methyl phenyl carbinol was present in traces (Table 2, run 5). Moreover, we expected that, for these specific reactions, high amounts of aldehyde were detrimental for the selectivity of the process, having the aldehyde the unique role of initiating the radical chain by generating PINO radical via molecule-induced homolysis.

Table 2. Aerobic oxidation of alkyl aromatics catalyzed by NHPI in the presence of CH_3CHO

Run	R-H	CH_3CHO (%)	t(h)	Conv. (%)	Selectivity (%)		
1[a]	CU	100	72	37	CH: -	CA: -	AP: 100
2[b]	CU	100	6	35	CH: 63	CA: 35	AP: 2
3[b]	CU	10	6	36	CH: 81	CA: 18	AP: 1
4[a]	ETB	100	5	48	PEHP: -	MPC: 8	AP: 85
5[c]	ETB	100	6	36	PEHP: 57	MPC: 6	AP: 37
6[c]	ETB	10	6	26	PEHP: 82	MPC: -	AP: 18

[a][42]; [b][43]; [c][44].

Indeed, our hypotheses were confirmed and we succeeded in increasing the selectivity in hydroperoxides up to values higher than 80% (Table 2, runs 3 and 6).

Both aliphatic and aromatic aldehydes were used to promote the oxidation of CU (Table 3). When operating in the presence of benzaldehyde, lower conversions were observed, probably due to the slower initiation step for aromatic aldehydes (Table 3, run 2), while other primary aldehydes like propionaldehyde (EtCHO, run 3) acted similarly to MeCHO.

As observed in the introduction, branched aliphatic aldehydes are usually more effective in promoting co-oxidative processes, probably due to their higher electronrich character with respect to primary ones, so that they are more easily oxidized. In fact, by operating in the presence of the secondary 2-ethylbutyraldehyde (Table 3, run 4) we were able to obtain analogous conversions and selectivity in just 2 h. With tertiary pivalaldehyde conversions were even higher, but with poor selectivity in CH (Table 3, run 5).

Nevertheless, from an industrial application point of view, acetaldehyde still remains the aldehyde of choice due to its low commercial value and the limited mass loss. Even if the presence of a polar solvent is crucial to maintain the polar catalyst in solution, it was possible to optimize the reaction conditions by increasing the concentration of the alkyl aromatic.

Thus, it was possible to conduct the aerobic oxidation of cumene at 70 °C in the presence of 1% NHPI, 2 % of acetaldehyde and with a volume ratio CU/CH_3CN of 5/2, achieving the desired hydroperoxide in 28% yield with 84% of selectivity.

Similarly, ETB was oxidized to PEPH with a lower yield (13%), but a higher selectivity (91%), by operating at the same temperature with 2% NHPI, 2 % of acetaldehyde and with a volume ratio ETB/CH_3CN of 1/1. These results are objects of two patent applications [45].

Table 3. Effect of different aldehydes in the aerobic oxidation of alkyl aromatics catalyzed by NHPI

Run	Aldehyde	t(h)	Conv. (%)	Selectivity (%)		
				CH	CA	AP
1	Me-CHO	6	36	81	18	1
2	Ph-CHO	6	25	93	7	-
3	Et-CHO	6	41	85	14	1
4	$(Et)_2CH$-CHO	2	40	86	14	-
5	$(Me)_3C$-CHO	2	62	58	37	5

Conclusion

The results reported in this chapter clearly demonstrate that the development of aerobic co-oxidative processes in the presence of aldehydes as sacrificial reductants is a field of great interest, especially for the design of new systems to perform selective epoxidations.

The reasons which push the scientific community to continuously increase the research activity in this field rely on the final task of progressively replace the traditional processes, based on oxidants like high-valent metal oxides and their mineral salts, with environmentally friendly and economic protocols operating with molecular oxygen as final oxidant.

N-hydroxyphthalimide is able to play a key role for the promotion of selective aerobic oxidations under very mild conditions. The combination of this versatile and economic catalyst with the approach of sacrificial co-oxidation allows the development of metal-free methods, which can find ample application in many fields of industrial relevance

REFERENCES

[1] Suresh, A. K., Sharma, M. M., Sridhar, T. T. *Ind. Eng. Chem. Res.*, 2000, 39, 3958; (b) Simandi, L. L. *Dioxygen Activation and Homogeneous Catalytic Oxidation*, Elsevier, Amsterdam, 1991; (c) Parshall, G. W., Ittel, S. D. *Homogeneous Catalysis*, 2nd edn, John Wiley and Sons, Inc., New York,1992; (d) Barton, D. H. R., Martell, A. E., Sawyer, D. T. *The Activation of Dioxygen and Homogeneous Catalytic Oxidation*, Plenum, New York, 1993; (e) Sheldon, R. A., Kochi, J. K. *Metal-catalysed Oxidations of Organic Compounds*, Academic, New York, 1981; (f) Bäckvall, J.-E. (ed.) *Modern Oxidation Methods*, Wiley-VCH Verlag GmbH, Weinheim, 2005.

[2] Jorissen, W. P.; van der Beek, P. A. A. *Rec. Trav. Chim.* 1926, 45, 245-247.

[3] (a) Ikawa, T.; Fukushima, T.; Muto, M.; Yanagihara, T. *Can. J. Chem.* 1966, 44, 1817-1825; (b) Tsuchiya, F.; Ikawa, T. *Can. J. Chem.* 1969, 47, 3191-3197.

[4] G. Franz, R. Sheldon, Ullmann's Encyclopedia of Industrial Chemistry, Oxidation, 4.4. *Homolytic Oxidation in the Liquid Phase*, Wiley-VCH, Weinheim, 2002, page 261.

[5] (a) Takai, T.; Hata, E.; Yamada, T.; Mukaiyama, T. *Bull. Chem. Soc. Jpn.* 1991, 64, 2513-2518; (b) Yamada, T.; Takai, T.; Rhode, O.; Mukaiyama, T. *Bull. Chem. Soc. Jpn.* 1991, 64, 2109-2117; (c) Mukaiyama, T.; Yamada, T. *Bull. Chem. Soc. Jpn.* 1995, 68, 17-35;

[6] Nagata, T.; Imagawa, K.; Yamada, T.; Mukaiyama, T. *Chem. Lett.* 1994, 1259-1262; (b) Yamada, T.; Imagawa, K.; Nagata, T.; Mukaiyama, T. *Bull. Chem. Soc. Jpn.* 1994, 67, 2248-2256.

[7] Dell'Anna, M. M.; Mastrorilli, P.; Nobile, C. F.; Lopez, L. *J. Mol. Catal. A: Chem.* 1996, 111, 33-36; (b) Iqbal, J.; Das, B. C. *Tetrahedron Lett.* 1997, 38, 1235-1238; (c) Wentzel, B. B.; Leinonen, S.-M.; Thomson, S.; Sherrington, D. C.; Feiters, M. C.; Nolte, R. J. M. *J. Chem. Soc., Perkin Trans.* 1, 2000, 3428–3431.

[8] (a) Bouhlel, E.; Laszlo, P.; Levart, M.; Montaufier, M.-T.; Singh, G. P. *Tetrahedron Lett.* 1993, 34, 1123-1126; (b) Dhar, D.; Koltypin, Y.; Gedanken, A.; Chandrasekaran, S. *Catal. Lett.* 2003, 86, 197-200; (c) Lim, S. Y.; Kang, M.; Kim, J.; Lee, I.-M. *Bull. Korean Chem. Soc.* 2005, 26, 887-891.

[9] Serra, A.; Rocha Gonsalves, A. M. d'A. *Tetrahedron Lett.* 2011, 52, 3489-3491.

[10] Serra, A. C.; Marçalo, E. C.; Rocha Gonsalves, A. M. d'A. *J. Mol. Catal. A. Chem.* 2004, 215, 17-21.

[11] Dolphin, D.; Traylor, T. G.; Xie, L. Y. *Acc. Chem. Res.* 1997, 30, 251-259.

[12] Zhou, X.-T.; Ji, H.-B.; Xu, J.-C.; Pei, L.-X.; Wang, L.-F.; Yao, X.-D. *Tetrahedron Lett.* 2007, 48, 2691–2695.

[13] Zhou, X.-T.; Tang, Q.-H.; Ji, H.-B. *Tetrahedron Lett.* 2009, 50, 6601-6605.

[14] Liu, C.; Shen, D.-M.; Chen, Q.-Y. *Eur. J. Org. Chem.* 2006, 2703-2706.

[15] Qi, J.-Y.; Li, Y.-M.; Zhou, Z.-Y.; Che, C.-M.; Yeung, C.-H.; Chan, A. S. C. *Adv. Synth. Catal.* 2005, 347, 45-49.
[16] Lehtinen, C.; Brunow, G. *Org. Proc. Res. Dev.* 1999, 3, 101-103.
[17] Mizuno, N.; Weiner, H.; Finke R. G. *J. Mol. Catal. A: Chem.* 1996, 114, 15–28.
[18] Punta, C.; Rector, C. L.; Porter, N. A. *Chem. Res. Toxicol.* 2005, 18, 349-356.
[19] Köckritz, A; Blumenstein, M.; Martin A. *Eur. J. Lipid Sci. Technol.* 2008, 110, 581-586.
[20] Venkateshwar Rao, T.; Sain, B.; Kumar, K.; Murthy, P. S.; Prasada Rao, T. S. R.; Joshi, G. C. *Synth. Commun.* 1998, 28, 319.
[21] (a)Mastrorilli, P.; Nobile, C. F. *Tetrahedron Lett.* 1994, 35, 4193; (b) Giannandrea, R.; Mastrorilli, P.; Nobile, C. F.; Suranna, G. P. *J. Mol. Catal. A* 1995, 103, 17; (c) Dell'Anna, M. M.; Mastrorilli, P.; Nobile, C. F. *J. Mol. Catal. A* 1996, 108, 57; (d) Song, G.; Wang, F.; Zhang, H.; Lu, X.; Wang, C. *Synth. Commun.* 1998, 28, 2783.
[22] Dell'Anna, M. M.; Mastrorilli, P.; Nobile, C. F.; Taurino, M. R.; Calò, V.; Nacci, A. *J. Mol. Catal. A* 2000, 151, 61-69.
[23] Babich, I. V.; Moulijn, J. A. *Fuel* 2003, 82, 607.
[24] Murata, S.; Murata, K.; Kidena, K.; Nomura, M. *Energy and Fuels* 2004, 18, 116-121.
[25] Ciccio, S. R.; Latronico, M.; Mastrorilli, P.; Suranna, G. P.; Nobile, C. F. *J. Mol. Catal. A* 2001, 165, 135-140.
[26] Fernández, I.; Pedro, J. R.; Roselló, A. L.; Ruiz, R.; Castro, I.; Ottenwaelder, X.; Journaux, Y. *Eur. J. Org. Chem.* 2001, 1235-1247.
[27] Theyssen, N.; Leitner, W. *Chem. Commun.* 2002, 410-411.
[28] Theyssen, N.; Hou, Z.; Leitner, W. *Chem. Eur. J.* 2006, *12*, 3401-3409.
[29] For recent reviews in the field, see: (a) Punta, C.; Gambarotti, C. In: *Ideas in Chemistry and Molecular Sciences: Advances in Synthetic Chemistry*, (Ed.: B. Pignataro), Wiley-VCH Verlag GmbH and Co. KGaA, Weinheim, 2010, pp 3-24; (b) Recupero, F.; Punta, C. Chem. Rev. 2007, 107, 3800-3842; (c) Minisci, F.; Punta, C.; Recupero, F. *J. Mol. Catal. A: Chem.* 2006, 251, 129-149; (d) Ishii, Y.; Sakaguchi, S. *Catal. Today* 2006, 117, 105 –113; (e) Ishii, Y.; Sakaguchi, S. in: *Modern Oxidation Methods*, (Ed.: J.-E. Bäckvall), Wiley-VCH Verlag GmbH and Co. KGaA, Weinheim, 2005, pp 119 – 164; (f) Ishii, Y.; Sakaguchi, S.; Iwahama, T. *Adv. Synth. Catal.* 2001, 343, 393-427; (g) Sheldon, R. A.; Arends, I. W. C. E. *Adv. Synth. Catal.* 2004, 346, 1051-1071; (h) Galli, C.; Gentili, P.; Lanzalunga, O. *Angew. Chem. Int. Ed.* 2008, 47, 4790-4796; (i) Coseri, S. *Catal. Rev.* 2009, 51:2, 218-292; j) Coseri S. *Mini-Rev. Org. Chem.* 2008, 5, 222-227.
[30] (a)Punta, C., Fontana, F., Pedulli, G. F. *Chem. Commum.* 2002, 688-689. (b) Amorati, R.; Lucarini, M. Mugnaini, V.; Pedulli, G. F.; Minisci, F.; Recupero, F.; Fontana, F.; Astolfi, P. Greci, L. *J. Org. Chem.* 2003, 68, 1747 – 1754.
[31] Minisci, F.; Recupero, F.; Cecchetto, A.; Gambarotti, C.; Punta, C.; Faletti, R.; Paganelli, R.; Pedulli, G. F. *Eur. J. Org. Chem.* 2004, 109-119.
[32] Howard, J. A. In: *Peroxyl Radicals*; Alfassi, Z., Ed.; Wiley and Sons: Chichester, 1997.
[33] Koshino, N.; Cai, Y.; Espenson, J. H. *J. Phys. Chem. A* 2003, 107, 4262.
[34] Ueda, C.; Noyama, M.; Ohmori, H.; Masui, M. *Chem. Pharm. Bull.* 1987, 35, 1372.
[35] Hermans, I.; Vereecken, L.; Jacobs, P. A.; Peeters, J. *Chem. Commun.* 2004, 1140-1141.

[36] (a)Minisci, F.; Zhao, L.; Fontana, F.; Bravo, A. *Tetrahedron Lett.* 1995, 36, 1697; (b) Bravo, A.; Fontana, F.; Fronza, G.; Minisci, F.; Serri, A. *Tetrahedron Lett.* 1995, 36, 6945; (c) Bravo, A.; Fontana, F.; Fronza, G.; Minisci, F.; Zhao, L. *J. Org. Chem.* 1998, 63, 254.
[37] Bravo, A.; Bjørsvik, H. R.; Fontana, F.; Minisci, F.; Serri, A. *J. Org. Chem.* 1996, 61, 9409.
[38] Minisci, F.; Gambarotti, C.; Pierini, M.; Porta, O.; Punta, C.; Recupero, F.; Lucarini, M.; Mugnaini, V. *Tetrahedron Lett.* 2006, 47, 1421-1424.
[39] Punta, C.; Moscatelli, D.; Porta, O.; Minisci, F.; Gambarotti, C.; Lucarini, M. In: *Mechanisms in Homogeneous and Heterogeneous Epoxidation Catalysis*, (Ed.: S.T. Oyama), Elsevier, Amsterdam, The Netherlands, 2008, pp 217 – 229.
[40] Spaccini, R.; Liguori, L.; Punta, C.; Bjørsvik, H.-R. *ChemSusChem* 2012, 5, 261-265.
[41] (a)Bjørsvik, H.-R.; Liguori L. WO 2007058544, 2007, pp. 1– 38; (b) Liguori, L.; Bjørsvik, H.-R. *Org. Process. Res. Dev.* 2011, 5, 997-1009.
[42] Einhorn, C.; Einhorn, J.; Marcadal, C.; Pierre, J.-L. *Chem. Commun.* 1997, 447-448.
[43] Melone, L.; Gambarotti, C.; Prosperini, S.; Pastori, N.; Recupero, F.; Punta, C. *Adv. Synth. Catal.* 2011, 353, 147-154.
[44] Melone, L.; Prosperini, S.; Gambarotti, C.; Pastori, N.; Recupero, F.; Punta, C. *J. Mol. Cat. A* 2012, 355, 155- 160.
[45] (a)Minisci, F.; Porta, O.; Recupero, F.; Punta, C.; Gambarotti, C.; Pierini, M. WO2008037435, 2008; b) Minisci, F.; Porta, O.; Recupero, F.; Punta, C.; Gambarotti, C.; Spaccini, R. WO2009115275, 2009.

In: New Developments in Aldehydes Research
Editors: L. Torrioni and E. Pescasseroli

ISBN: 978-1-62417-090-4

Chapter 6

SYNTHESIS AND STRUCTURE OF GOSSYPOL CONDENSATION BIS-PRODUCT WITH 2-AMINO-4,6-DIOXYPYRIMIDINE IN ACIDIC ENVIRONMENT

K. Z. Tilyabaev*, F. G. Kamaev, A. M. Yuldashev, and B. T. Ibragimov

Institute of Bioorganic Chemistry, Uzbek Academy of Sciences, Uzbekistan

ABSTRACT

The reaction of gossypol with 2-amino-4,6-dioxypyrimidine in acidic environment has been studied.

The structure of obtained product has been investigated by UV-, IR- and ^{1}H NMR-spectroscopy methods. Obtained results suggest that gossypol with 2-amino-4,6-dioxypyrimidine in acidic environment forms not C-N but C-C product due to keto-enol and lactam-lactim tautomerism of the latter.

Keywords: Gossypol, 2-amino-4,6-dioxypyrimidine, acidic environment, synthesis, structure, keto-enol and lactim-lactam tautomerism

INTRODUCTION

Gossypol (*I*) (Figure 1) is triterpene aldehyde of cottonseed, which exhibits unique biological properties [1]. Based on aldehyde derivatives of *I* with primary heterocyclic amines as well as those with heterocyclic compounds containing active CH_2- groups some effective medical preparations have been developed [2,3].Systematic investigations [4,5] have demonstrated that the introduction of heterocyclic fragment-containing substituents to the

* Email: tilyabaevzaid@mail.ru.

molecule *I* allows obtaining compounds with low toxicity and retained therapeutic effects such as antiviral immunomodulating and others, which is explained by their interferone-inducing action. In connection to this, the synthesis of gossypol (*I*) derivatives containing heterocyclic substituents is of a high interest. The purpose of this work is to study the reaction of *I* with 2-amino-4,6-dioxypyrimidine (*II*) (Figure 1) and the structure of the product obtained. *II* has been chosen because pyrimidine derivatives are known to possess wide ranged biological activities allowing to use them as pesticides, herbicides and intermediate substances for the preparation of medicines [6]; thus, the combination of these properties of *II* with physiological activity of *I* can be perspective.

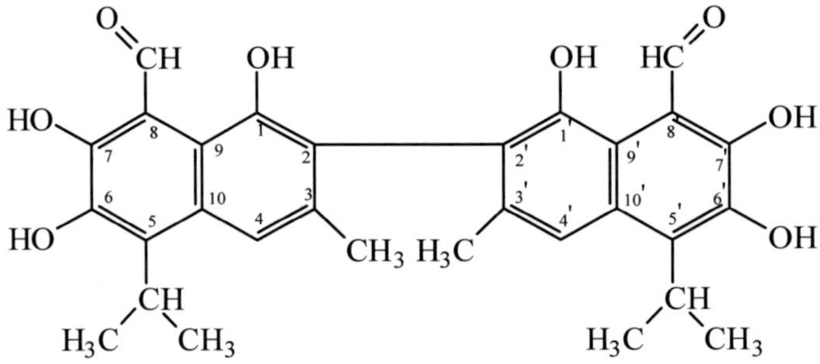

(I)

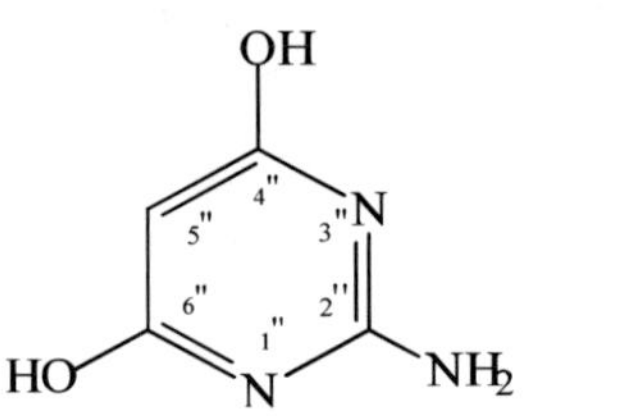

(II)

Figure 1. Structures of gossypol (I) and 2-amino-4,6-dioxypyrimidine (II).

Results and Discussion

By the reaction of *I* with *II* the formation of corresponding Schiff base (*III*) was expected. The scheme of this reaction is given on Figure 2.

Due to insolubility of *II* in water and common organic solvents it was dissolved in concentrated $HCl_{(aq)}$, and compound *I* was dissolved in 96% EtOH. The product as wine red precipitate was obtained by the combination of afore-mentioned solutions at room temperature. The structure of the product was studied by UV-, IR- and ^{1}H NMR-spectroscopic methods.

It has to be noted that obtained spectral data do not confirm the structure of *III*. In ^{1}H NMR-spectrum there is no signal of CH-proton in substituent fragment of *III*, however, there is a signal at 10.81 ppm, which is typical for condensation products of *I* with compounds containing active hydrogen atoms [3].

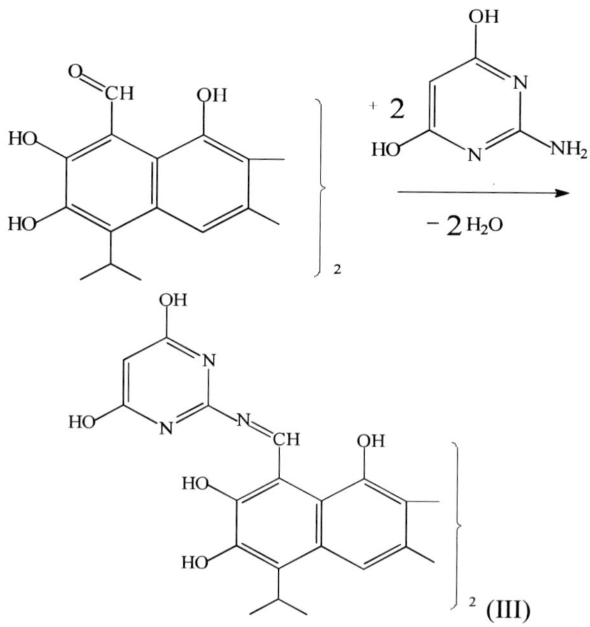

(III)

Figure 2. Scheme of the formation of gossypol bis-Schiff base (III) with 2-amino-4,6-dioxypyrimidine.

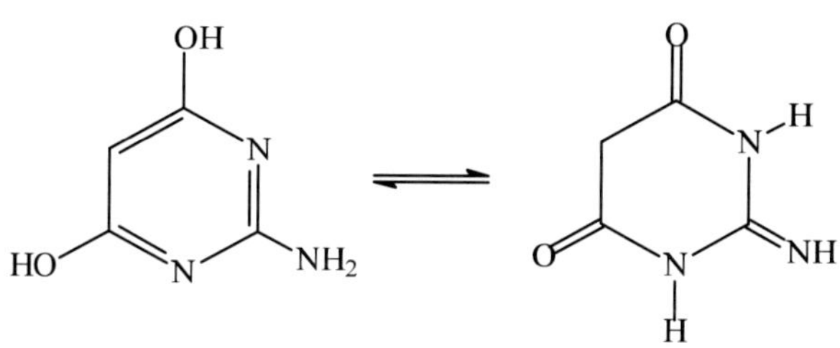

(IV)

Figure 3. Keto enol and lactim-lactam tautomerism of 2-amino-4,6-dioxypyrimidine (lactam tautomer–IV).

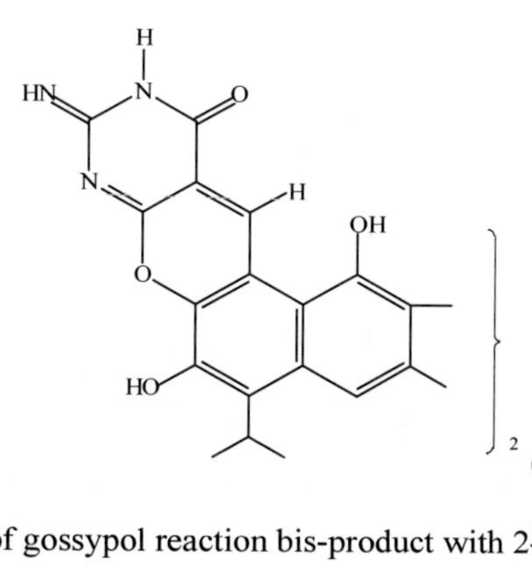

(V)

Figure 4. Suggested structure of gossypol reaction bis-product with 2-amino-4,6-dioxypyrimidine in acidic environment.

Taking into consideration that substituted oxypyrimidines undergo keto-enol and lactim-lactam tautomerism, the possible shift of tautomeric equilibrium for *II* in acidic environment (HCl) towards lactam tautomer (*IV*) (Figure 3), can cause a new reactive center (similar to that in the reaction of *I* with barbituric acid [3]) to appear and therefore, the product, which is different from the expected one, can form.

The above-mentioned facts allow suggesting the following structure for the reaction bis-product of *I* with *II* in acidic environment (*V*, Figure 4.)

^{1}H NMR-spectral data also allow suggesting that the intramolecular dehydration resulted in a heterocycle joining the gossypol backbone with the substituent (the absence of the C7–OH resonance upfield (14.5–16.0 ppm) and the absence of one of the substituent NH protons (Figure 4)).

The presence in the IR-spectrum of an absorption band at 1670 cm^{-1} characteristic of C=C bonds whose π electrons were conjugated with lone pairs of electrons of oxygen atoms [7] also supports this assumption.

EXPERIMENTAL PART

Compound *I* was prepared at the experimental plant of the Institute of Bioorganic Chemistry, Uzbek Academy of Sciences, from side products of cotton oil production. Chemical reagents and solvents were obtained from Aldrich.

The melting point of *V* was measured in a glass capillary. TLC was carried out on Silufol UV-254 plates with the development in an iodine chamber.

The UV spectrum was registered on a Specord spectrophotometer. The IR spectrum was recorded on an IR-10 spectrophotometer in the region of 3700–750 cm^{-1} in KBr tablets. The ^{1}H NMR spectrum (DMSO-d_6) was registered on NMR-UNITY 400+ spectrometer with a working frequency of 400 MHz and tetramethylsilane as an internal standard.

V:1,1',6,6'-Tetrahydroxy-9,9'-Diimino-5,5'-Diisopropyl-3,3'-Dimethyl-9,9'10,10'-Tetrahydrogen-11H,11'H-2,2'-Dibenzo[5,6]Chromeno[2,3d] Pyrimidin-11,11'-Dione

5 mL of 30% $HCl_{(aq)}$ solution containing 26 mg (0.0002 mol) of *II* was added to 15mL of 96% ethanol solution containing 52 mg (0.0001 mol) of *I* at room temperature and stirred under the same conditions for 1 hour. Course of reaction has been controlled by using TLC method on Silufol UV- 254 plates eluting with ethyl acetate. The reaction mixture was concentrated by evaporation under reduced pressure and the residue was washed several times with minimal volume of 96% ethanol. M.p. >350°C. R_f=0.17 (Silufol UV-254, ethylacetate). Yield- 85%. Found, %: C 65.01; H 4.48; N 11.90. $C_{38}H_{32}N_6O_8$. Calculated, %: C 65.14; H 4.57; N 12.00.

UV-spectrum (EtOH), λ_{max}, nm, (lgε): 235(4.61), 519 (4.19)

IR-spectrum (KBr), ν, cm^{-1}: 3400-3600 (O-H, N-H), 1670 (C=C), 1610 (C=O)

^{1}H NMR-spectrum: (400 MHz, DMSO-d_6), δ, ppm (*J*, Hz): 1.58 (6H, d, J=6.1, -CH(C<u>H</u>$_3$)$_2$); 2.10 (3H, s, C*3*-CH$_3$); 4.19 (1H, m, J=6.1, -C<u>H</u>(CH$_3$)$_2$); 8.02 (1H, s, H*4*); 8.67 (1H, s, C-OH); 9.51 (1H, s, C-OH); 9.65 (1H,b.s, >C=NH); 10.62 (1H, b.s, -C(O)-NH); 10.81 (1H, s, C-H in a chromene fragment)

CONCLUSION

Obtained results suggest that gossypol with 2-amino-4,6-dioxypyrimidine in acidic environment forms not C-N but C-C product via inter- and intramolecular dehydration. This chapter was prepared by invitation from Nova Science Publishers Inc.

REFERENCES

[1] Dodou, K. Investigations on gossypol: past and present developments. *Expert Opin.Investig.Drugs,* 2005, 14(11), 1419-1434

[2] Saiitkulov, A.M., Ershov, F.I., Aslanov, Kh.A., Tazupakhova, E.B., Maulianov, S.A., Achilova, G.Sh., Amanova, S.Kh. Relation between interferon-inducing activity of natural interferon inducers and their chemical structure. *Antibiot. Khimioter.,* 1991, 36(7), 39-42.

[3] Baram, N.I., Ismailov A.I. Biological activity of gossypol and its derivatives. *Chem. Nat. Compd.,* 1993, 29(3), 275–287.

[4] Baram, N.I., Ismailov, A.I., Biktemirov, L., Paizieva, R.Z., and Ziyaev, Kh.L. Proizvodnyye gossipola i ikh fiziologicheskaya aktivnost (Gossypol derivatives and their physiological activity), in Problemy i perspektivy razvitiya khimii prirodnykh i fiziologicheski aktivnykh veshchestv (Problems and Prospects of Development of Chemistry of Natural and Physiologically Active Compounds), Tashkent: Fan, 1988, pp. 78–99.

[5] Ziyaev Kh.L. Azotsoderzhashchiye proizvodnyye gossipola i sozdaniye lekarstvennykh preparatov na ikh osnove (Nitrogen-containing derivatives of gossypol and development of medical preparations on their basis). Doctoral thesis. Tashkent. Institute of Bioorganic Chemistry. 2009, pp. 1-39.

[6] Qing, W., Xuwu, Y., Shengli, G., Qizhen, S. Preparation and the Standard Enthalpy of Formation of 2-Amino-4,6-dimethoxypyrimidine and the Related Complexes of Copper. *Chem. Pap.,* 2003, 57 (2), 97 101.

[7] Katritskii, A.R., Embler, A.P., Fizicheskie Metody v Khimii Geterotsiklicheskikh Soedinenii (Physical Methods in Chemistry of Heterocyclic Compounds), Moscow: Khimiya, 1966, pp. 546-548.

Index

D

N

O

P

Q

R

S

T